DEMOCRACY IN A HOTTER TIME

DEMOCRACY IN A HOTTER TIME

Climate Change and Democratic Transformation

EDITED BY DAVID W. ORR
FOREWORD BY BILL MCKIBBEN
AFTERWORD BY KIM STANLEY ROBINSON

The MIT Press
Cambridge, Massachusetts
London, England

The MIT Press would like to thank the anonymous peer reviewers who provided comments on drafts of this book. The generous work of academic experts is essential for establishing the authority and quality of our publications. We acknowledge with gratitude the contributions of these otherwise uncredited readers.

This book was set in Adobe Garamond Pro and Berthold Akzidenz Grotesk by Westchester Publishing Services. Printed and bound in the United States of America.

Library of Congress Cataloging-in-Publication Data

Names: Orr, David W., 1944– editor. | McKibben, Bill, other. | Robinson, Kim Stanley, other.
Title: Democracy in a hotter time : climate change and democratic transformation / edited by David W. Orr ; foreword by Bill McKibben ; afterword by Kim Stanley. Robinson.
Description: Cambridge, Massachusetts ; London, England : The MIT Press, [2023] | Includes bibliographical references and index.
Identifiers: LCCN 2022059067 (print) | LCCN 2022059068 (ebook) | ISBN 9780262048590 (paperback) | ISBN 9780262376471 (epub) | ISBN 9780262376464 (pdf)
Subjects: LCSH: Democracy and environmentalism.
Classification: LCC JC423 .D439867 2023 (print) | LCC JC423 (ebook) | DDC 321.8028/6—dc23/eng/20230228
LC record available at https://lccn.loc.gov/2022059067
LC ebook record available at https://lccn.loc.gov/2022059068

10 9 8 7 6 5 4 3 2 1

To Gus Speth

Contents

Foreword

Bill McKibben

Not long ago some of us began to form a group for older Americans who wanted to work for progressive change (Third Act, it's called). As we thought about how to target out efforts, we kept coming back to the same idea, one shared by this remarkable book: we'd all taken for granted both the physical stability of our planet and the stability of our democracy. And taking those things for granted had been extraordinary mistakes.

Climate change first. In an earlier era, we'd imagined pollution as something going wrong at the edges: the Cuyahoga River burning, Los Angeles shrouded in smog. Within two years of the first Earth Day, Congress had passed the Clean Air and Clean Water Acts, and indeed the air and water had begun to get cleaner (amazing how fast politics worked in those days). But now we face a truly existential challenge that goes to the very core of our economy. Some of us remember those pictures that came back from the Apollo missions, but Earth doesn't look like that anymore; half the sea ice in the Arctic has melted, trading white ice for blue sea.

And democracy. Those of us who lived through, say, Watergate understood that the system could be stressed, but we never imagined mobs of Americans attacking police officers with flag poles in the nation's capitol to stop the counting of votes.

In both cases, the damage built up incrementally over decades, and indeed they were closely related. The drive of American conservatism in the years after the Clean Air Act to stifle regulation of pollution eventually produced, say, the Supreme Court's 2022 ruling that largely gutted that law.

Along the way it also produced the long list of sad practices—over-the-top gerrymandering, dark money, voter suppression—that threaten our democracy in the most profound ways.

But there's good news too: the necessary steps to limit the climate crisis (above all the rapid deployment of solar and wind power) could help us in the democracy fight as well. We're reminded by Putin's invasion of Ukraine that control over the scarce supplies of hydrocarbons can go hand in hand with regressive autocracy. (Americans should know this too, since our biggest oil barons, the Koch Brothers, have done more than anyone to deform our democracy.) Sun and wind are available everywhere; as we move in their direction, we can move toward more localized control over our destiny.

It's going to be a very close call, however—both of these crises are timed tests. We may have just a few elections between us and a failed state, and just a few years before the level of carbon in the atmosphere kicks off changes that will never be reversed. So now is precisely the moment to think deeply and clearly about these questions, which the authors of this volume surely do. It's at the intersection of these fights—for a stable climate and a stable democracy—that the future will be determined.

INTRODUCTION

David W. Orr

In 1770, Tom Paine, thirty-three, was teaching school in Lewes, England; newly married Thomas Jefferson was building Monticello; and nineteen-year-old James Madison was a student at Princeton. The convergence of ideas, people, circumstance, and serendipity we call the American Revolution was still in the future. By 1800—thirty years later—these men had written some of the most brilliant reflections on government ever. The Colonies had declared their independence, won a war against the mightiest army in Europe, conceived a new constitutional order, launched a bold experiment in large-scale democracy, elected George Washington as the first president, and peacefully transferred power from one faction to another. Against all odds, they had imagined and launched the first modern democracy. Imperfect though it was, the fledgling nation had the capacity for self-repair evolving toward "a more perfect union." Sojourner Truth, in that year of 1800, was three years old. Our challenge, similarly, requires us to begin the world anew, conceiving and building a fair, decent, and effective democracy, this time better fitted to a planet with an ecosphere.

This book is a scouting expedition to that possible future and a speculative inquiry about the transition to a more durable, fair, and resilient democracy, and what that will require of us. We are close either to a precipice or to a historic turning point, and for a brief time, the choice is ours to make. But let's begin with where we are now.

In the summer of 2022, temperature in London reached 40°C (104°F); record heat, drought, and fire scorched 60 percent of Europe and much of

China; New Delhi registered 45°C. Water levels in Lake Mead dropped to 27 percent of capacity and are still falling. The Great Salt Lake was shrinking fast. But much of Pakistan was flooded by unprecedented monsoon rains. The frequency and intensity of heat waves and heavy rainfall worldwide continued to increase. In dry regions, drought and wildfire intensified. The Greenland ice sheet melted faster than ever recorded. The century-long rise in global sea level was accelerating, and ocean temperatures continued to increase.[1] The National Oceanic and Atmospheric Administration recorded 419 ppm CO_2 in the atmosphere.[2] Changing reality made it necessary to invent words like "pyrocumulonimbus" to describe never-before-seen things such as the massive columns of smoke and heat rising from western US fires into the stratosphere, creating their own weather systems below. All of this and more was happening with a global warming of 1.9°F (1.1°C) above the late-nineteenth-century average, exceeding the highest temperatures recorded in at least two thousand and possibly more than a hundred thousand years. Even assuming current policies to restrain emissions are fully implemented, we are heading toward a possible 5.4°F (3°C) rise by the end of the century, double the 1.5°C (2.7°F) red-line of "dangerous interference with the climate system" set at the UN Climate Change Conference, COP21, in 2015. After that, who knows.[3] We do know, however, that at some point along the gradient of rising temperature, everything changes. First sporadically, as now, then as a constantly shifting "new normal," and finally as a series of self-amplifying runaway cascades.[4] There is no safe haven anywhere from effects of rising heat on ecosystems, societies, economies, political systems, and even our own health and morale.[5] In Naomi Klein's words, it is "an everything issue" unprecedented in its velocity, scale, and duration.[6] It is also an everywhere issue. I wish—we all wish—it were otherwise, but it is not.

On the other hand, there are reasons to hope that long-overdue change is finally happening. In August, Congress passed the first major climate legislation in US history.[7] The costs of renewable energy and improved efficiency continue to decline and are increasingly competitive with energy from fossil fuels and nuclear power almost everywhere.[8] Sizeable majorities of the public support action on climate change and adoption of renewable energy.

Business and finance are moving mostly in the right direction because the liabilities of "green" investments are lower and profits higher.[9] Buildings and entire cities are being designed to be carbon neutral, driven by market demand, better technology, superior design, and more comprehensive building standards and international codes. And New York and two other states have amended their constitutions to include the right to "clean air and water and a healthful environment." Whether all this is too little, too late, time will tell. Had we acted earlier, the hole we've dug would not be nearly so deep, but even after the first authoritative warnings decades ago, we kept digging. It did not have to be this way.[10]

By the mid-1980s there was more than enough scientific evidence for the United States to lead a worldwide transition to energy efficiency, renewable energy, and ecologically smarter design of economies, cities, transportation, farms, and factories. We were warned, repeatedly, in ever greater detail, but did not act, committing "the greatest dereliction of civic responsibility in the history of the Republic."[11] In other words, we squandered whatever margin of safety we might once have had. Our failure to meet the challenge early on, when it would have been much easier, cannot be excused by a lack of technology or even by economics, since efficiency, renewable energy, and superior design have long been cheaper, faster, and more resilient than the alternatives and without the incalculable costs of climate chaos.[12] The cause, rather, is political. Our fossil-fuel- and corporate-dominated democracy seems to have stalled out; our institutions corrupted by too much unaccountable money and elected officials with too much ambition and too little integrity; our various media by too much venom, too little concern for the common good. Now, in an ongoing right-wing insurrection, we are vexed and troubled, still struggling to solve even the most basic problems, including climate change, that threaten our own survival.

In short, we face two related existential crises: a global crisis of rapid climate change and potentially lethal threats to democracy. We believe these are related, and because people have an unalienable and hard-won right to choose how they are governed and to what ends, democracy is worth fighting for. We believe, further, that a more robust democracy would be a more effective, fair, and durable way to organize the transition to a post-fossil-fuel world than any possible alternative. And there's the rub: democracy as it exists may not survive for long on a rapidly warming Earth, but on the other hand, as James Hansen says, "We cannot fix the climate until we first fix democracy." Fixing democracy, however, requires fundamental improvements that, among other things, protect the written and unwritten rules that contain our political disputes and provide greater political equality, economic justice, and protection of the rights of future generations. It also requires improving government and governance by calibrating law, regulation, policy, taxation, administration, and behavior to how the Earth works, as a complex physical system with feedback loops and long gaps between causes and effects. Our predicament is compounded because some fraction of CO_2 remains in the atmosphere for more than a millennium, rendering our plight a "long emergency" measured in the time required to stabilize the climate system and restore the Earth's energy balance.[13] Accordingly, we should study long-lived institutions, cultures, economies, and political systems for what we might learn about how to render our own more durable, decent, and fair over the long haul. Effective responses to both crises will require systems thinking, long time horizons, and the capacity to "solve for pattern."[14]

Rapid climate change, in short, "presents the most profound challenge ever to have confronted human social, political and economic systems."[15] As such, it is first and foremost a political and moral crisis, not one solely of technology or economics, as important as those obviously are.

The climate crisis comes at a particularly bad time. Authoritarianism is advancing here and elsewhere.[16] Authoritarian governments sometimes move faster than democracies but have a dismal record on climate, environment, and human rights issues. Governments run solely by experts might possibly deploy technological and scientific expertise more surely than democracies, but they would have no monopoly on wisdom about where and how to apply technology for what purpose, or when to stop.[17] For these and many other reasons, we will bet on "we the people" and our collective capacity to strengthen, expand, and reinvent the institutions of democracy at all levels in the time available to meet new challenges posed by a warming climate.[18] But it won't be easy.

Novelist Amitav Ghosh puts it this way: "Climate change represents, in its very nature, an unresolvable problem for modern nations in terms of their biopolitical mission and the practices of governance that are associated with it."[19] Modern democracy, in particular, grew out of what historian Walter Prescott Webb once described as "the Great Frontier," the opening of the Americas to European migration.[20] The changed ratios of people to land, minerals, and forests for a time relieved demographic stress in an overcrowded Europe. The resulting affluence, he thought, made democracy possible by reducing scarcity and improving quality of life. Increased wealth sanded down some of our rougher edges, but the sheer rush of pell-mell capitalism also gave license for the genocide of Native peoples, slavery, Jim Crow laws, inequality, and ecological ruin. Our "biopolitics," in other words, made climate change a predictable outcome of a system built on exploitation of people and nature alike and powered by fossil fuels. But as the ratio of natural resources to people becomes tighter, the struggles over the fair distribution of what's left will become more bitter and eventually could be fatal to democracy.[21]

Our predicament is rather like an engine failure on a fire truck speeding to a five-alarm emergency. Even as we address the global bonfire driving climate chaos, we will also have to divert our attention to repair the machinery by which we do the public business of voting, legislating, administering, taxing, subsidizing, regulating, and judging. But the machinery of government, rickety though it was, did not break down. It was sabotaged. The neoliberal movement, spawned by Frederick Hayek, Milton Friedman, and others in the late 1940s, proposed to throw sand in the gears and deflate the tires of the New Deal "administrative state" and then denounce government as hopelessly inefficient and markets magically otherwise. They intended to replace government built in large part as a countervailing power to offset that of robber barons, rogue capitalists, and footloose corporations with that of that mythical never-seen creature called a free market. Their success required both Ayn Rand-ian zealotry and amnesia—the great forgetting of the darker aspects of American history that needed sunlight and healing. They succeeded all too well in shrinking our imagination about what good government could be and what it could do while making short-term market share the lodestar of public policy and law. Lewis Powell's 1971 memo, circulated in the US Chamber of Commerce, called for a corporate counterattack against the New Deal and progressive policies, including civil rights, environmental protection, social equity, public health, education, political accountability, and transparency.[22] Conservative foundations, and others, spent lavishly to create "think tanks" to give a patina of legitimacy to libertarianism and protect use of "dark" money in election campaigns as a form of free speech.[23] They spent even more to create a right-wing media universe centered on FOX "news," whose business plan aims to keep people tuned in by distraction and anger, appealing to the part of the brain neuroscientists call the amygdala—the ancient reptilian brainstem where ghouls of fear and inchoate violence lurk in the shadows.[24] The predictable results included a flood of unaccountable money, growing inequality, gerrymandered electoral districts, restrictions on voting rights, and an eruption of cynicism, mendacity, and nihilism, all of which exacerbated a widening gap between rural and urban voters that in turn helped to elect a rogue president who, among his other offenses, organized a coup to overturn a legitimate election and undermine the electoral system. If that were not enough, a theocratic majority on the Supreme Court, supported by a militant Christian nationalist movement, intends to impose its pre-Enlightenment predilections on twenty-first-century Americans.[25] The war against American democracy could not have happened at a worse time.

All the while, carbon emissions are rapidly changing Earth into a different and less hospitable planet for humans. Even in the rosiest scenarios imaginable, a warming climate driving more capricious weather will destabilize governments and increase conflicts over water, food, and land, stressing global supply chains and international institutions to the breaking point.[26] It could be worse, but without concerted preventative action, it is not likely to be better. In either case, governing will become more difficult at all levels because of increasing climate-driven weather disasters, the difficulty of making systemic solutions necessary to manage multiple problems without causing new ones, and intensifying conflicts between rich and poor.

One thing more: the US Constitution rigorously protects private property but not what we hold in common and in trust, such as climate stability and biological diversity. It does not acknowledge our dependence on ecological systems with complex feedback loops and cause and effect separated in space and time. It does not protect future generations who will live with the consequences we leave behind—Jefferson's "remote tyranny," across generations.

In sum, there is no plausible resolution for the convergence of crises in the "long emergency" that does not include healing our uncivil civic culture and reforming our politics, policies, governing institutions, and laws to accord with Earth systems and a larger sense of solidarity and self-interest that includes our posterity.[27] Our best and, I believe, our only authentic hope is in a renewed commitment to repair and fundamentally improve democratic institutions and governments at all levels. It won't be easy to do, but much easier than not doing it. Democracy has always demanded a great deal from citizens. Now it requires learning how to be dual citizens in a political system *and* in an ecological community and knowing why these are inseparable. We must learn—perhaps relearn—the arts of tolerance, neighborliness, ecological competence, and the kind of patriotism that shifts loyalties from "I," "me," and "mine" to "we," "ours," and "us," including posterity and other species. I imagine that in some future time we will be judged not just by our technological prowess, but by our skill in the arts of effective, wise, and accountable government—a democracy undergirded by a civically smarter and more supportive citizenry and provisioned by a better-thought-out and more carefully designed technology in an economy harmonized to the carrying capacity of Earth's various ecosystems and grounded in vibrant and diverse and resilient local communities.[28]

Yale political scientist Hélène Landemore, here and in her book *Open Democracy*, argues persuasively that we have underestimated our collective intelligence and political competence. She argues that good reasons exist to extend the scope and reach of democracy and find better ways to educate and engage citizens. With Congress deadlocked, the Republican Party mired in quicksand, a politicized Supreme Court, and overburdened government agencies, it is a very good time to enlist the creativity, talents, patriotism, and practical skills of 333 million Americans whose common future is in jeopardy. Writing in 1919, W. E. B. Du Bois put it this way: "The real argument for democracy is . . . that in the people we have the source of that endless life and unbounded wisdom which the rulers of men must have, . . . a mighty reservoir of experience, knowledge, beauty, love, and deed."[29] The challenge is how to harness the great power and intelligence latent in that untapped reservoir to build a just, inclusive, and sustainable world powered by sunlight.

Visions are easy to dream but hard to implement. If that better, more inclusive, and effective democracy is to grow and flourish, and if we are to be reconciled to the Earth, those more expansive and necessary visions must live in the minds and lives of our youth. For that reason, among others, educational institutions are on the front lines in the battle for democracy and a habitable Earth. Every graduate from every school, college, or university should know how the Earth works as a physical system. They should understand the civic foundations of democracy. They should come into adulthood with a sense of authentic hope in a world still rich in possibilities. For those who teach and administer, it is time to ask, what is education for, especially now? It is time to rethink the enterprise called "research" and better deploy our intelligence and compassion to meet human needs for food, shelter, health care, education, conviviality, safety, and energy.

* * *

Finally, a word about what this book is and is not: it is a "scouting expedition" intended to describe some of the salient features of the topography ahead. There are four features that we dare not ignore. First, it is imperative to "disinvent fire" and make a rapid transition to efficiency, renewable energy, and better design of nearly everything.[30] Second, we must reckon with the limits posed by the topography of the centuries ahead. As acknowledged in John Wesley Powell's 1879 proposal to tailor settlement in the "arid regions" of the US West to the fact of water scarcity, we will confront a series of unmovable limits imposed by climate, ecology, water, thermodynamics, and our own fallibilities.[31] One way or another decisions will be made about the scale of the human enterprise relative to the ecosphere, the just distribution of costs and benefits within and across generations, and what we owe to posterity.[32] Third, the transition ahead is both political, having to do with "who gets what, when, and how," and moral, having to do with matters of fairness and decency. I see no plausible way to reach that better future without significantly improving our politics, what Vaclav Havel calls "living in truth."[33] Finally, we must better understand ourselves and what we've become shaped by, a culture of consumption that is ravaging the ecosphere, and what we might yet become, with foresight and a bit of help from the angels of our better nature.[34]

This is not, however, primarily a book about policy or recent developments in energy technology or the sins of capitalism, as important as those are. The focus is upstream, on the political and governmental institutions where decisions about policy, technology, and the economy are made, or not.[35] It is, rather, a conjecture from various perspectives about the human response to rapid climate destabilization, and possibilities for improving democratic institutions and civic culture to meet the stresses ahead. Implicit throughout are questions of whether democracy can survive through the turbulent years ahead and become an asset in a transition to a future much better than that in prospect. And our focus is mostly on the United States, in large part because of its greater influence in causing the problem and its greater potential to launch the systemic changes necessary to a decent human future.

The contributors worked at the intersection of a conundrum that sets short-term, fragmented, incremental changes against the need for faster systemic change. We know how to do things in small steps, but the goal should be to take small steps that lead to systemic change with as little disruption as possible.[36] They also worked at the intersection of a five-alarm crisis and an engrained habit of "generally not getting overly excited about anything," what Naomi Klein calls "the fetish of centrism."[37] That creates a related communications conundrum. While bombs were falling on London in 1940, for example, Winston Churchill offered the British people only "blood, toil, tears, and sweat," not a lecture on the joys of urban renewal. On the other hand, Jack Nicholson's character Colonel Nathan Jessup in *A Few Good Men* famously blurted out to a tense courtroom, "You can't handle the truth." It all depends. In tough situations, Mark Twain advised that one should tell the truth because it will amaze your friends and confound your enemies. In our circumstances, however, how do we tell the truth without also inducing despair and fatalism? But even in the best scenarios, the temperature of the Earth will rise steadily in the foreseeable future and the resulting weather will be increasingly chaotic, causing immense suffering and losses that did not have to be.

Michael Oppenheimer's overview of the facts about climate change and long-term implications of a warming Earth sets the stage for the chapters that follow, including his "and yet" expression of faith that we will rise to the challenge. Part I deals with the complexities of democracy and its potential. Frances Moore Lappé argues that unseen possibilities exist "as we bust the myths about democracy being out of reach," if we can summon the imagination and courage to do "what we thought we could not." Hélène Landemore focuses on the prospects for "open democracy" that build on our capacities as citizens working in better-designed forums and formats. Daniel Lindvall addresses the capacity of democracies to protect the right of future generations to a habitable Earth. In other words, we have possibilities grounded in our history, and in our capacities for learning, creativity, and empathy.

Part II examines a few of the many roadblocks on the road to democratic renewal and new challenges, of which there are many. Some of these are long-standing structural and procedural problems embedded in the Constitution and our history. Some owe to the very nature of politics in a democracy, which is to say tribalism and human cussedness. William Barber's opening calls us to seize the moment to transform not only how our society is powered but how we deploy political and economic power to "secure a future for us all, . . . the best that our collective humanity has to offer." For a society buffeted by uncontrolled technology, David Guston proposes "no innovation without representation," reminiscent of the commonsense precaution to "look before you leap." The point is that the market alone is a bad way to deploy complex technologies that affect health, environment, and civility in ways that we often fail to anticipate. Holly Buck's essay focuses on the "outrage-industrial complex" that has "supercharged" our animosities and warped our "conceptions of democracy." The solutions require us to understand and restructure "our media environment . . . [to] support our common humanity." Finally, Fritz Mayer addresses the thorny issues that plague the politics of climate change in an anarchic world of sovereign states where consensus and self-interest collide.

Part III focuses on issues of policy and law. Bill Becker's analysis of American democracy indicates that we have broken through gridlock before. This time, however, is more demanding: we must surmount "narrow tribalism, manufactured outrage, the absence of a unifying national vision, and the loss of fundamental values." Lincoln's "angels of our better nature" wait in the wings. Ann Florini, Gordon LaForge, and Anne-Marie Slaughter propose to better deploy nongovernmental organizations, corporations, finance, and business, building on the imaginative use of self-interest beyond the realm of government to advance the public good. Katrina Kuh and James May analyze the "constitutional silence on the environment" and "judicial abdication" on matters of environment and climate—that is, human survival. Remedy will require an "open-eyed reckoning with how and why the constitutional status quo is failing" and, presumably, jurists and judges who understand the relationships among Earth systems science, jurisprudence, and the human prospect. Batting cleanup in this part, Stan Cox assesses the state of our ignorance relative to what's knowable and what's not on a warming planet where the climate is "becoming less predictable year by year." The prognosis, if not quite desperate, is not quite not desperate, either. "We and future generations," he writes, "will face the need for profound adaptation" much of which will depend on work at the local level in neighborhoods and communities, which is, I think, good news.

Finally, because our difficulties and perplexities mostly begin with how we think and what we think about, part IV deals with the effort to improve thinking through that complicated process, education. Michael Crow, president of Arizona State University, and his coauthor William Dabars describe the "fifth wave" university response to climate change and the necessary combination of learning, innovation, and forbearance essential for a decent future. The five-alarm nature of climate chaos requires revising curriculum, research, and innovation throughout higher education and changing requirements for graduation so that every student in every field knows what planet they're on, how it works, and why such things are important for our public life and for their own lives and careers. Wellington ("Duke") Reiter's chapter describes the Ten Across initiative, which joins the ten major cities along US Highway 10 from Jacksonville to Los Angeles as a "proving ground for the most critical issues of our time." It is also an example of the creativity and innovation possible in well-led and imaginative universities. Finally, Richard Louv, author of *Last Child in the Woods* and the founder of Children and Nature Network, describes the importance of the experience of nature early in childhood and how that opens the democratic vista rooted in a "deep emotional attachment to the nature around us." The sense of connections early in childhood, by which I mean the awareness that we are kin to all that ever was, is now, and ever will be is the emotional bedrock for a democratic order and those otherwise elusive habits of heart that defy mere reason.

A scouting expedition does not yield a detailed map with GPS precision. But, like the explorations of John Wesley Powell in the Southwest or the Lewis and Clark Expedition (1804–1806), we've covered what we consider to be the most important features of the complex, chaotic, and surprising world ahead. There is, however, much more to be said. Here we've emphasized the need to join a lucid understanding of biophysical reality with how we do the public business fairly, competently, democratically, and with foresight.

THE CHALLENGE

Michael Oppenheimer

I wonder how Svante Arrhenius who, in 1896, published the first estimate of how much Earth would warm if carbon dioxide levels in the atmosphere doubled because of coal burning, would react if he could witness the twenty-first century's rapidly rising concern over climate change. By historian Spencer Weart's recounting, Arrhenius, a Swede who won the 1903 Nobel Prize in chemistry for an entirely different discovery—the ionic theory of solutions—had welcomed the idea of a warmer Sweden.[1] Apparently, Arrhenius believed humans could solve any problems arising from a warming he reckoned was thousands of years in the future.

What Arrhenius couldn't foresee was the exponential economic expansion of the twentieth century and the accompanying growth of emissions of carbon dioxide and other heat-trapping greenhouse gases. At the time, other scientists thought that emitted carbon dioxide would rapidly dissolve in the ocean (eliminating the warming problem entirely) rather than understanding that much of it was destined to remain airborne for millennia—science that began to emerge sixty years later.[2] If Arrhenius and his contemporaries had considered the possibility that, eventually, removing the human-caused excess of carbon dioxide to prevent a dangerous warming would receive serious consideration by scientists and policy experts, they might have assumed that the technology to do so would be readily available, another by-product of the same phase of industrial development that caused emissions to grow in the first place. If all this had occurred across a thousand years, maybe they would have been right.

Regrettably, the human inflation of Earth's natural greenhouse effect has occurred at a pace ten times faster than Arrhenius foresaw and, at this juncture, has outpaced the widespread implementation of solutions to the problems it has created. The rapid growth of atmospheric carbon dioxide, the most important human-made greenhouse gas, began to worry scientists only a few years after monitoring of carbon dioxide atop Mauna Loa volcano in Hawai'i began in 1957. These measurements provided the single most important piece of evidence that the greenhouse effect was about to turn into the greenhouse problem and, potentially, a climate change catastrophe. Not entirely coincidentally, the carbon dioxide buildup became obvious just a few decades sooner than the emergence of reliable, safe, *and* affordable renewable energy options. However, that small difference in timing is precisely why we find ourselves at such risk today.

In addition to the phenomenal rate of increase of industrialization relying largely on fossil fuel combustion, two additional facts explain how we arrived at the current treacherous situation: once emitted, carbon dioxide has accumulated in the atmosphere, because natural removal by dissolving in the ocean is slow, leaving us, and future generations, with a legacy problem; and the greenhouse problem has not and will not fix itself by exhaustion of fossil fuels or by reaching some natural plateau in the warming. Only completely eliminating emissions of the major human-made greenhouse gases, and perhaps deploying artificial means to remove some of the aforementioned legacy emissions from the atmosphere, will solve the problem short of widespread catastrophe. This brief chapter lays out the scientific reality of what we face if we fail to do so and asks the question, "Are we up to the task?"

The ongoing increase in emissions and resulting warming has occurred despite climate change having risen high on the global policy agenda since the signing of the UN Framework Convention on Climate Change at the 1992 Earth Summit at Rio de Janeiro. This treaty, and the concerns that led to it, produced a set of institutions to encourage international cooperation to reduce emissions and prepare for some degree of unavoidable climate change. In turn, these developments bore fruit in the form of the Kyoto Protocol and the Paris Agreement, among other diplomatic milestones, but nevertheless have fallen far short of the aspiration spelled out in Article 2 of the Framework Convention—avoiding "dangerous anthropogenic interference with the climate system."[3] This goal was concretized in the Paris Agreement as limiting greenhouse gas emissions to an extent "consistent with holding the increase in the global average temperature to well below 2°C above preindustrial levels and pursuing efforts to limit the temperature increase to 1.5°C above preindustrial levels."[4] These levels were based on a combination of feasibility and scientific understanding of the looming risks, like loss of a significant chunk of the Antarctic ice sheet, a possibility entailing a sea level rise exceeding one meter in this century and ten to twenty times as much in subsequent centuries.[5]

Recent events have highlighted both the possibility and the difficulty of reaching the Paris Agreement's goal. On the one hand, the US government, having reawakened from a climate stupor during the Trump administration, has authorized implementation of a package of financial commitments aimed (mostly) at decarbonizing the US economy nearly fast enough to meet the Paris targets, assuming all other countries with significant greenhouse gas emissions act in step.

On the other hand, the disruptions in global and especially European energy supplies resulting from Russia's war on Ukraine have given both coal and oil renewed hope of stretching out their historical dominance in fueling the world's economies. Simultaneously, the COVID-19 pandemic, whose direct effects on emissions were large in the near term but will eventually fade (as opposed to the effects of the Russia-Ukraine war, which may be long-lasting), drained away some of the political momentum from the climate change issue. This should serve as a strong reminder to governments that climate change won't be the only mega-issue on their plates as it is gradually managed and hopefully contained over the rest of this century. Any long-term solution to the climate problem must be robust against surprising political and economic as well as scientific developments.

Meanwhile, if emissions mitigation efforts continue to move along unevenly in some key emitter countries, and barely at all in others, we will find ourselves ever-more focused on the climate changes occurring right before our eyes, the even greater climate threats looming closer and closer, and the concomitant need to adapt *flexibly* in response. I emphasize the need for flexibility because the level of scientific uncertainty about the details of climate change becomes larger the farther into the future we project beyond midcentury. Uncertainties come in two varieties: things we need to know better about the physics of the atmosphere, ocean, land, and ice in order to project future climate changes along any plausible emissions pathway; and how our technological, economic, social, legal, and political (e.g., policy) behavior over time will determine which one of an infinite number of potential global emissions pathways the world will follow. None of this is easy. Climate physics is complicated enough, while forecasting human preferences and behavior accurately over multiple decades is nigh impossible.

For example, the key to understanding the overall amount of global warming from a doubling of greenhouse gases in the atmosphere is called the *climate sensitivity*. The Intergovernmental Panel on Climate Change (IPCC) estimates that it is likely to lie between 2.5°C and 4°C.[6] When this quantity was first reported more than forty years ago, the estimate was 1.5°C–4.5°C. Just eight years ago, the estimate was still 1.5°C–4.5°C, although with considerably more confidence than when the same estimate was made originally. This shows important progress, but also that progress is uneven and sometimes very slow. In this case, that is because the uncertainty in sensitivity is determined by the physics of clouds and how they change with warming, a very difficult problem to solve.

Clouds produce a feedback on the warming from the greenhouse gases, meaning that how they change in response to the greenhouse gas buildup can either amplify or reduce the gases' initial warming effect. It turns out the clouds can do both—increases (decreases) in some types of clouds at certain locations amplify (reduce) warming. Increases or decreases in other types of clouds do the opposite. At this time, we simply don't have enough observations of clouds under varying conditions, or sufficient understanding of some of the basic physics of cloudiness, to narrow the climate sensitivity further.

Taken together, feedbacks in the climate system are very likely positive, reinforcing warming. This includes, along with the response of clouds, processes such as the amount of additional heat trapping due to water vapor fed into the atmosphere by the warming ocean surface. However, how far the net effect of all such feedbacks lies on the positive side of zero is uncertain. Another key uncertainty is the potential response of the Antarctic ice sheet to warming, an uncertainty that complicates estimation of sea level rise late in this century.

Some level of uncertainty about the physics of the climate will always be with us. Nevertheless, we need to make plans for both emission mitigation and adaptation right now, without being paralyzed by uncertainty. That is why flexible approaches aimed at multiple objectives are so important—they spread the downside risk of a misjudgment about the amount and rate of future warming. Decisions on emission reduction should produce other benefits, like reduced air pollution, even if our climate projections are off the mark. Adaptation actions should benefit the life and livelihoods of those who need help most, regardless of whether they ever wind up in the bull's-eye of a climate catastrophe.

EVENTS AND TRENDS

Climate change presents itself as long-term trends, such as global warming itself or the increasing duration of decades-long regional climate phenomena, like the drought now occurring in the US West. The greenhouse gas buildup also affects short-term phenomena, causing changes in the location, frequency, duration, and intensity of individual day-to-day weather events including heat waves, storms with extreme precipitation totals, and very dry or very humid days. All manner of troubles follows from both events and trends, but at the current time, there is increasing focus by both policy-makers and scientists on the individual events. This is partly because monetized harms from the most extreme events, that is, those which greatly diverge from what is typical, are causing increasing amounts of damage, both in the United States and worldwide. The cost and salience of these events—like the intense rains and flooding accompanying Hurricane Harvey that deluged Houston in 2017, the level of the storm tide that contributed to the $20 billion in damages to New York City from Hurricane Sandy in 2012, or the approximately forty thousand excess deaths in the 2003 heat wave in central western Europe, and the suspected role of climate change in ginning up each—has uncovered a shocking lack of preparation in some countries, whether rich or poor.

Furthermore, beginning about twenty years ago, scientists developed methods for distinguishing events that had become more extreme because of the greenhouse gas buildup—hotter or wetter or drier—from events that might have occurred anyway in the historical climate. Recently, ways to assign a cost to these excess damages have emerged. This way of tracking the "smoking gun" bears obvious implications for apportioning responsibility in a very tangible way, whether in a judicial courtroom or the court of global public opinion.

Finally, trends are by their very nature long-term phenomena—in forecasting, it generally takes thirty years or more of data to label a change as a trend. People have an opportunity to anticipate and accommodate or otherwise adapt to trends, although sometimes while experiencing considerable pain and loss, financial and otherwise. Some may move out of harm's way, like farmers experiencing extended drought or people living in low-lying coastal areas who may pick up and move. Others may pressure their governments to build protections and adaptations like the flood defenses surrounding New Orleans or the irrigation and drinking water supply systems that allow Arizonans to thrive. Extreme events are quite another matter because they are, by definition, rare.

When a once-per-century event (e.g., the so-called hundred-year flood, the flood with a probability of occurrence, in any one year, of 0.01) occurs, people notice because it disrupts their lives, sometimes costs them a lot of money to fix the damage, and might kill family members, friends, and neighbors. When two such "rare" events occur within a month in close proximity, as happened when two Category 5 hurricanes struck the Leeward Islands and Puerto Rico in 2017, people begin to wonder whether climate change is at play; scientists try their best to get an answer. Increasingly, they can, in terms of what the probability is of the event occurring in today's world, and what it would have been had greenhouse gas levels never increased via human activity. Or, from another perspective, how intense the event would have been in the counterfactual world versus how intense the event actually was.

PACKING THEM IN, ONE AFTER ANOTHER

Given the effects of climate change, we expect the most intense tropical cyclones (the generic name for hurricanes, typhoons, and cyclones—which are the same phenomena, occurring in different parts of the world) to become yet stronger, although the details of where and how much remain sketchy. However, we are *very confident* in what changes to expect in occurrence of very hot days and sequences of such days (heat waves): in the future, historically rare heat events associated with disastrous outcomes will occur more and more frequently at most locations.[7] Heat waves like those that occurred in India and Pakistan in early summer of 2022, or western North America in 2021 (with a reported mortality total of over 1,400 excess deaths in the US Pacific Northwest and southwest Canada), or, as noted above, west-central Europe in 2003, will become more common and progressively more intense.[8] Similarly, destructive flooding from intense rainfall of the sort that devastated parts of Kentucky and South Korea in summer 2022 or Northern Germany in summer 2021, as well as regional drought resembling that experienced in the US West over the past two decades, will occur increasingly frequently. Flooding from coastal storms will spread farther inland to higher elevation because of sea level rise: at many coastal locations worldwide, the historical once-per-century flood level will be reached *once per year* by 2050, even if the Paris Agreement's 2°C target is achieved.[9]

Examining changes in extreme events, not just trends in averages, reveals the full scope of the problem. The world has warmed nearly 1.2°C compared with the late nineteenth century. Yet heat levels that were, back then, attained only once every fifty years in the world's continental areas as a whole now occur once every ten years and will return nearly once a year if we allow global average temperature to reach 4°C.[10] Furthermore, periods of relief will gradually vanish in summer at many places, as hotter and hotter days cluster closer and closer together during the hot season. Even if we manage to restrain global warming to below the Paris Agreement targets of 1.5°C or 2°C, this sort of amplification of the frequency and clustering of extreme events will occur, although much more moderately than at 4°C.

A key measure of success at adaptation will become not just how effectively we recover and minimize damages from individual events, but whether the job at hand can be fully or only partially completed before the next extreme event of a similar or even entirely different type delivers another knockout punch at the same location, or one nearby, or even a distant location that is connected politically or economically to the first. Such *compound* or *connected events,* which scientists are now examining in detail, include repeated heat waves with little or no relief in between, which present a particular threat to the very old and very young, those with preexisting conditions, and those who cannot afford air conditioning or, as occurred in the US Pacific Northwest in 2021, those who expected never to need it. Another example under study involves the conditions caused by destructive hurricane strikes followed by intense heat waves, a combination likely to increase as the world warms. The effects of tropical cyclones often include loss of electric power and thus interruption of air conditioning and loss of any possible relief from the subsequent heat. Yet another compounding of damage occurs when wildfires or other climate-related disasters occur during a period of non-climatic stress on our health and emergency response systems, like COVID-19.

Where compound events become the norm, full recovery may never be attained. The issue then becomes whether our ability to respond is up to the challenge. I use the word "we" cautiously because the approximately eight billion individuals on this Earth possess wildly different capabilities to adapt. The question arises, of course, whether those with larger resources will attempt to cut the others loose and restrict protection to themselves, or whether an effort aimed more at the collective well-being will come to pass. Consider the responses to increasing climate risk and its manifestations in terms of actual or threatened loss of life, livelihood, or home. One such response is mobility, which includes actions along a spectrum of agency ranging from planned, voluntary migration to forced displacement.[11] Climate change can result in both, but global estimates of how many people might move as the climate changes are not especially reliable or useful.

Instead, consider the episodic migration away from unstable political and economic conditions in areas that also experience disastrous climate events with some regularity—for example, from Central America northward, or both within and outward from the Mediterranean basin, which stretches from North Africa through the Middle East and around to the Iberian Peninsula. Climate change will exacerbate the very conditions that sometimes enhance such migration flows, providing one of many links that will keep those who have access to resources from decoupling from those who do not.

The moral quandary over what is just and equitable reaches yet deeper. New research suggests that international migrants will tend to move preferentially to destination countries with lower climate change risk than their point of origin.[12] The effect of increased stringency of border policy is to increase the exposure of potential migrants to dangerous climate changes by preventing their emigration without allowing them any agency in the choice. In addition, it is well established that those at the lowest end of the income spectrum who might otherwise desire to migrate are more likely to be stuck where they are or, at the very least, inhibited from moving because of their relative impoverishment. Sadly, for many, their impoverishment will be deepened by climate change, leading to a spiral wherein more and more people in some countries are actually less and less able to adapt to climate change by emigrating. So, while emigration from countries suffering high levels of climate damage may increase, so may the size of populations who desire to move but cannot.

Nevertheless, bear in mind that most migration occurs *within* rather than between countries, and unequal access to resources to facilitate adaptation is a big problem within wealthy as well as poor countries, as the ineffective and racially tinged preparation, evacuation, and recovery efforts leading up to, during, and after Hurricane Katrina striking New Orleans in 2005 so amply and so sadly illustrated.

CONCLUDING THOUGHTS

Asking governments and individuals to solve the climate problem is asking a lot. Problems like this don't really ever get solved in the usual sense of arriving at a situation where attention can turn away entirely and forever. The climate problem is more like managing foreign relations with other countries, some friendly, some not, and with relations in perpetual flux. In the case of climate change, however, much of the evolving risk will come not just from what other countries are doing or not doing by way of a solution but also from the way nature works, and our imperfect understanding of it. Events will occur and phenomena will emerge that we had not thought of and don't initially understand, or that we had discarded as unimportant or so unlikely as to be dismissed, which nevertheless turn out to pose serious threats. Our understanding will change over time, perhaps radically. We'll need to prepare for good breaks and bad ones and have scenarios for action, whether via emissions mitigation, adaptation, or some third possibility that can be deployed rapidly, effectively, efficiently, and equitably. The latter might include concepts that are currently viewed by many, myself included, as unwise, like geoengineering or a largely nuclear-powered world, or unreasonably optimistic, like 100 percent reliance on renewable energy, globally, within the 2050 timeframe.

Doing all this while fending off other threats is a tall order for institutional arrangements that are often ineffective and sometimes corrupt, and are distrusted accordingly. I will never forget watching Hurricane Katrina play out, people dying needlessly because of ineffective defenses and faulty emergency planning and rescue operations that were reckless in their insensitivity to peoples' differing situations and capacities. Similar examples can be cited worldwide. These failures are not so much about individual misjudgment as about lack of institutional perception of emerging challenges along with weak-kneed political leadership.

AND YET . . .

Ingenuity is a large part of what makes us human. We make it a habit to give technologies free rein until, all too often, their risks are not just apparent but deadly. Sometimes the technologies become so unmovable (or unremovable) that they are like political incumbents—determining *our* future, rather than vice versa. But we try, and often succeed at regaining control and managing the risks back in line with the benefits we reap. Again, the use of "we" is problematic—for many communities, even for whole countries, that balance may never be achieved if we don't deal with climate change globally, comprehensively, and immediately.

The passing of Mikhail Gorbachev as war rages in Ukraine is a reminder that our creativity released a two-faced genie—we eventually contained but did not extinguish the threat of a large-scale nuclear exchange, stuffing it back in its bottle but not quite screwing the cap back on. The situation requires attention and management, forever. And so it will go with climate change, I believe, at least for the foreseeable future. The energy revolution is real, and we should take heart in that and take every opportunity to spur it on, make every effort to beat the Paris Agreement's goals.

Carbon-free technology, however, is just the beginning. It will soon be or already is at hand. The other part of what makes humans human—the social, economic, and political relations, attitudes, and arrangements we develop, the processes by which we reduce emissions and manage climate change in order to minimize its impacts on humanity—are of greater importance to the outcome. Organizing at multiple scales—ranging from supranational groups down to the neighborhood level—initiatives that are public or private, collective or individual, that resonate with each other more than they interfere, can generate radical change. Powerful forces like the oil industry will eventually die or totally transform by capitalizing on solving the problem. Newly powerful forces and new institutions will emerge, some also malign, but if you believe in progress over the long haul then, in retrospect, climate change will fit the pattern of humans creating severe problems, then (sometimes) managing them away from disaster. Just as with nuclear arms, we let climate risk run out of control for too long. But just as with arms control, we are learning fast how to rein it in.

I DEMOCRACY?

1 DEMOCRACY: IT'S WHO WE ARE AND WHAT THE EARTH NEEDS

Frances Moore Lappé

Democracy's decline, staggering economic inequality, and the climate crisis are frightening global developments that, if approached as distinct threats, can appear insurmountable. Solutions arise, however, as we ground ourselves in one insight: these challenges reflect interwoven threats that share a common root in systems of thought from which we must—and can—free ourselves.

But before digging into this unifying approach, let me share the learning journey that brought me to this realization.

Just out of college and with no clear direction in sight, it dawned on me that I might find my footing if I started with what's most basic. So, my big "ah-ha" was a four-letter word: "Food!" Because we all must eat to survive, I thought that if I could just grasp why so many were going hungry, I could unlock the mysteries of economics and politics and discover a meaningful path forward.

At the time, experts were shouting that the problem was scarcity. There's just not enough! Paul Ehrlich's *Population Bomb* so frightened my peers that some committed to remain childless.

Ah, I thought, I'll start there, asking, Is scarcity *truly* our problem?

Digging in at the UC Berkeley agricultural library, I quickly discovered—to my great surprise—that hunger is caused not by a scarcity of food but by a scarcity of democracy. By this I meant that humanity was producing enough calories for all, and since no one chooses to go hungry, the very existence of hunger proves that some of us are deprived of power, even the most basic power to eat. The root meaning of *power* is "to be able"; so it seemed that millions were being needlessly *dis*empowered.

Democracy, to me, meant—and still means—governance in which we each have a voice. Thus, in a true democracy, no one is powerless. Deprivation of life's essentials therefore became my proof of democracy's absence. Today, per person, the world produces about one-fifth more calories than fifty years ago, and more than enough for all.[1] Yet, almost four in ten of us still lack access to an adequate diet.[2]

Other evidence of the absence of democratic decision-making?

We have created a global food system so destructive and wasteful that 80 percent of agricultural land is devoted to producing livestock that return to us only 18 percent of our calories.[3] Our food system contributes as much as 37 percent of all greenhouse gas emissions, while greatly speeding habitat loss, species decimation, and, worldwide, hundreds of oceanic "dead zones," largely a result of nitrogen runoff from meat-centered agriculture.[4] Moreover, it has created suffering for farmworkers, almost half of whom are poisoned yearly, and inequity in control of farmland worldwide so extreme that 1 percent of farms control 70 percent of farmland.[5]

Given these life-defeating realities, it should be no surprise that only one-fifth of humanity lives in a democracy, and worldwide democracy is in decline.[6]

So, my questions continued to grow until I got to this puzzler: How could it be that we together are creating a world that no one of us individually would choose? After all, no one turns off the alarm in the morning eager to starve another child or heat the planet.

Soon, an answer began to emerge. Could it be that *Homo sapiens*' big brains come with a big hazard? They enable conceptual thinking and vivid imagination, which serve us, of course. But what if—as creatures of the mind—we humans, uniquely, create mental maps of the world that function as filters?

"It is theory which decides what we can observe," observed Albert Einstein.[7] Or, as attributed to both Plato and the Hopi Indians, "Those who tell the story rule society."[8]

Now, that's all well and good *if* our stories are life serving. But I came to see our culture's dominant and reductive mental map—increasingly globalized—as life destroying.

And what is that story?

It goes like this:

Sure, most humans are capable of goodness, but realistically all we can count on are three traits: we are selfish, materialistic, and competitive. Thus, most important is organizing economic life so that it is driven by these three universal impulses.

Then, assuming these limitations in our nature, economics relying on self-interest appear more central to society's well-being than democratic political governance. The latter is seen as a real stretch—assumed to be dependent on some degree of consideration of the whole and thus beyond the capacities of most of us. This assumption seems to have a particular grip on the American mind.

Then, prioritizing economic life, we celebrate what many call a "free market," presumably meaning free of rules set by what many view as inefficient and corruptible government. We become blind to the reality that no market is free. Ours, for example, is driven by one organizing principle: do what brings the highest return to capital, that is, existing wealth. So of course, wealth accrues to wealth.

It is no surprise, then, that today in the United States wealth is so tightly held that the three richest Americans control more wealth than the entire bottom half of the population.[9] And the pace of concentration has only accelerated in recent decades.[10] In the process, the United States has become a global outlier in income inequality as well, ranked by the World Bank as having more extreme concentration of income than in 120 nations.[11]

Corporate concentration has kept up with such skewed personal accumulation—defying the very premise of a competitive market economy. In the food industry, for example, as of 2021, just four or fewer firms had come to control at least half of the market for nearly 80 percent of our groceries; and just four companies now control over 80 percent of the beef industry.[12]

In agriculture, we cling to the notion of independent family farms, but over half of our farmland is rented.[13] And as of 2017, 4 percent of the largest US farms controlled almost 60 percent of farmland.[14] In 2021, Bill Gates became a symbol of continuing farmland consolidation as he became the largest US farmland owner, now with total holdings six hundred times the size of the average US farm.[15]

Farmers are also dependent on corporate buyers that are highly concentrated. A handful of powerful transnational companies dominate every link of the US food-supply chain: from seeds and fertilizers to slaughterhouses and cereals and beers to supermarkets.[16]

A narrowly profit-driven economy without democratic guideposts is not only grossly unfair, but it can also be dangerous. In the United States it has generated a food industry in which 60 percent of the calories we consume are ultra-processed and empty of nutrition. Our diet has thus become a factor contributing to most major health threats, from diabetes to cancer.[17] Red meat remains a mainstay of the American diet, even as the World Health Organization in 2015 deemed it a probable carcinogen and classified processed meat as a carcinogen.[18]

As noted, our grain-fed, meat-centered diet is also a climate threat. It requires vast acreage as well as agricultural chemicals and energy for production, processing, and transportation—all combining to generate more than third of all greenhouse gas emissions.[19] Plus, producing our grain-fed-meat diets contributes to rainforest destruction, depleting carbon stores. Grazing as well as feed production drive "more than two-thirds of the recorded habitat loss" in Brazil's Amazon and nearby regions, reports World Wildlife Fund.[20]

In all the above, the antidemocratic circle of control determining our future shrinks; and this tight grip of economic players translates into political power. Mega-firms and billionaires fill the campaign coffers of political candidates who, once elected, feel obligated to serve their funders.[21]

Examples of political funders with great influence are billionaires David Koch—who died in 2019—and his brother Charles. Since the 1970s, they have contributed $100 million to help birth the antigovernment Tea Party movement and to bolster the far-right wing of the Republican Party.[22] The

Kochs also backed the American Legislative Exchange Council, an organization of state lawmakers and business lobbyists that drafted "model legislation"—with six hundred bills making it into law by 2019—and coached lawmakers in reducing taxes and weakening environmental protections.[23] In 2022 the Koch organization Americans for Prosperity funded candidates in more than six dozen midterm state primaries.[24] Of course, monied interests with a range of political perspectives and motives help fund political campaigns.[25]

A money-driven political system robs citizens of their legislators' time, attention, and loyalty.

Those we elect to serve us spend 30 to 70 percent of their time "dialing for dollars" to keep their jobs.[26] Moreover, we permit lobbyists for monied interests—including, of course, the fossil fuel and agribusiness interests worsening the climate crisis—to twist the ears of officeholders. Today, there are nearly twenty lobbyists pressing their interests in Washington for every one legislator that we citizens have elected to represent us.[27]

A democracy driven by private wealth is democracy in name only.

BREAKING FREE OF DEADLY THOUGHT TRAPS

Crises can either paralyze or awaken. Since paralysis ensures eventual destruction of life on Earth, let's delve into what awakening might feel like by asking, Can we reverse our massive losses by rethinking democracy itself and humans' capacity for it?

Rethinking democracy could start with the obvious: "Elections plus a market economy" do not democracy make. Many nations have had both and remained both autocratic and unjust. Today, the United States has both, and yet its quality of democracy is ranked way down, coming in sixty-second worldwide, between Samoa and Panama among more than two hundred nations, judged by political rights and civil liberties. This global ranking is generated by Freedom House, established in the 1940s, with Eleanor Roosevelt as an early leader.[28]

Our sad status is painful news. But consider this: With what we know about the built-in deficits in America's democracy, would it not be vastly more discouraging if we ranked well? Appreciating our low status means that there are many nations from which we can learn.

ARE HUMANS EVEN CAPABLE OF DEMOCRACY?

Given today's multiple crises and democracy in retreat worldwide, many Americans may wonder whether humans are even capable of democracy—especially as we learn little about how the world's leading democracies function. Yet, confidence in our capacities for democracy are essential for action. So here is evidence that our species comes wired with essential traits for democratic living.

One: A deep sense of fairness In a clever experiment, even Capuchin monkeys reject a ration if they see their neighbor monkey getting one that looks tastier![29] And, many parents likely recall "It's not fair!" as one of their children's earliest protests.

Two: Empathy Even Adam Smith, sometimes considered the godfather of greed, wrote, "How selfish soever man may be supposed, there are evidently some principles in his nature, which interest him in the fortune of others, and render their happiness necessary to him, though he derives nothing from it except the pleasure of seeing it."[30] Empathy is made possible by our ability to see from another's perspective—a neat trick that even babies as young as seven months achieve.[31] We also find a hint of early empathy in the fact that babies cry at the sound of other babies crying but rarely at a recording of their own cries.[32] Or consider this: there's certainly no reason to think we humans are less empathetic than the rhesus monkey; in one experiment they refused food (in one case, for twelve days) if their eating triggered an electric shock hurting another monkey.[33] Also, notwithstanding the common assumption that crowds panic in emergencies, research suggests that in such moments we're more likely to die from our compassion—as we use precious time to help others—than from a competitive stampede to save ourselves.[34]

Three: Generosity From these first two traits, it should not surprise us that experiments have found that giving makes us happier than receiving.[35]

So, yes, humans evolved key traits that democracy both calls forth and depends on. Moreover, there is also a case to be made that *only* democracy can meet deep human needs—the physical and beyond. And what are these needs? Beyond food, water, and oxygen, our species thrives when three interdependent psychological needs are met as well:

One, power—a sense of agency in knowing we have a voice and are not mere pawns.

Two, meaning—feeling that our lives have significance beyond our own survival.

And three, connection—knowing we are not alone but are connected with others as we create lives of power and meaning.

The fulfillment of these needs captures what is meant by the enigmatic concept, "dignity." And, arguably, democracy is the only form of living in community that holds the possibility of embodying and nourishing the dignity of all.

Equally true, democracy is the only pathway enabling our species to become protectors and nourishers of planetary life, not—as we've long been—its destroyers. Why? History has proven that autocratic governance, by definition, is self-interested, focused on its own hold on power and thus unable to be accountable for the well-being of all, in the future as well as in the present.

LEARNING IN AND FOR DEMOCRACY

So how might educators nurture lifelong learners, able and eager to grasp this unique evolutionary moment, when humanity must quickly take huge steps toward sustainability requiring knowledge across many disciplines?

In creating learning environments that weave together specialized knowledge, in fields such as environmental science, attention must be given to developing one's effective voice to act on what one learns within a polity, especially in an era when elected democracy is under threat here and abroad.

That's a big order. It suggests the necessity of creating classroom environments in which learners are challenged to probe square-one questions about pursuing "public Welfare," the nation's purpose established in our constitution's preamble. And in the process are encouraged to

find one's own burning question and the question behind that question . . . and the next;
rethink democracy not only as an equitable structure of government but as a culture of often-unspoken assumptions;
explore these often-unspoken assumptions about human nature that both stand in the way and can free us into lifetimes of effective and rewarding action.

So here we are.

Progress is possible as we bust the myths about democracy being out of reach or simply a dull duty and begin to celebrate and nurture aspects of our species' nature that make democracy a rewarding way of life. We avert despair that stifles action as we encourage learners to tune into an historic development in this moment—a citizens' "movement of movements" for democracy itself. Perhaps for the first time in our history, environmentalists, social justice advocates, labor activists, and more are joining hands demanding key democracy reforms.[36]

In this movement, democracy is understood as not only a structure of government but also as a way of living in which we each have a voice. The movement will grow as our citizens learn from an early age that democracy is deeply rewarding—involving the thrill of exercising one's voice.

"POSSIBILISM" IS ALL WE HUMANS NEED

Given the powerful threats we face, perhaps optimism is hard to come by, so we may need an additional reframing. In our world of connection and continuous change, it is *not possible to know what's possible.* Thus, we can let go of disappointment or negative self-judgment when we feel unable to muster optimism.

All most humans need to act is a sense that it is *possible* our actions could make a difference. So let us cultivate "possibilism"—the assumption that our actions send out ripples large and small, and even our inaction changes the world around us, though not in ways aligned with our values.

To step into the unknown does also require what's difficult for most of us: courage. While much is uncertain, we *do* know that to save the life of our beautiful Earth requires doing what we thought we could not do. And, yes, stepping into the new can be frightening, but as Eleanor Roosevelt so aptly challenged us, "Do something every day that scares you."[37]

We are such social creatures that we take on the qualities of those close to us. So, now is the moment to seek friends who will challenge, encourage, and inspire us to do what we thought we could not do. As we reject despair and let go of finger wagging, we can discover democratic engagement as thrilling. It meets our deepest human needs for power, meaning, and connection.

And as we step out, we can take heart in the wisdom of the first African American federal judge, Justice William H. Hastie:

> Democracy is a process, not a static condition. It is becoming rather than being. It can easily be lost, but never is fully won. Its essence is eternal struggle.[38]

2 NO DECARBONIZATION WITHOUT DEMOCRATIZATION: TO SAVE THE CLIMATE, OPEN DEMOCRACY

Hélène Landemore

The planet is burning. The Intergovernmental Panel on Climate Change's warnings about the consequences of rising temperatures are becoming increasingly dire, and Russia's invasion of Ukraine has set off a race in Europe and elsewhere to achieve energy independence through rapid transformations of the economy.

With decarbonization becoming such an urgent priority, it is tempting to consider political shortcuts. Why not try enlightened despotism or a technocracy of experts, picking the best climate scientists and engineers and empowering them to make the decisions for us? Why not embrace the Chinese method of forcing through sweeping changes and swatting away any misguided resistance from below?

But in truth, there can be no decarbonization without democratization. Addressing climate change requires fighting off the technocratic temptation and instead deepening democratic decision-making at all levels—local, national, and global. It also requires extending democracy to the economic sphere, specifically to firms and organizations where climate-affecting decisions are made. The point is to include more people, interests, and ways of thinking in the relevant law and policy processes as well as in firm- and industry-level business decisions that affect the production of greenhouse gases. And, because of the urgency of climate change, democratization along those two dimensions—deepening and extending—must therefore happen quickly and decisively.

I begin by addressing the technocratic temptation of handing out power to experts and show why it is misguided. I then turn to the necessity to frame climate change as a problem of environmental justice. Next, I argue for democratizing our systems in a way that opens them up to the agency and wisdom of ordinary citizens, specifically through the use of randomly selected citizens' assemblies. Finally, I extend the reasoning to the economic sphere.

THE TECHNOCRATIC TEMPTATION

As urgent as solutions to climate change have become, so, too, has the need to address the growing disenchantment with democracy. Without rehearsing all the various indicators of democratic disillusionment—from unfavorable public sentiment to the rise of voter abstention and declining trust in elected politicians and public institutions—it is clear that many people now regard democracy as more of a problem than a solution.

If people do not trust their elected representatives to address fundamental policy issues such as national security, public health, education, and so forth, how would they ever trust them to manage something as massive, long-term, and multidimensional as climate change? A seemingly reasonable alternative is to turn to the same experts and scientists who have been warning about the dangers of climate change for several decades now.

The apparent advantages of a technocratic approach are twofold. First, in addition to knowing more than anyone else about climate change, climate scientists and adjacent experts would be able to follow the science, uninfluenced by vested interests, climate skeptics, ignorant and irrational voters, or social movements that resist the economic costs of change. Knowledgeable, efficient, and effective policies and laws to curb carbon emissions would be guaranteed.

Second, leaders who are unencumbered by mechanisms of democratic accountability can act swiftly and decisively. As we saw during the early stages of the COVID-19 pandemic, China's authoritarian government was able to control the virus by imposing massive quarantines on millions of people, building new hospitals in the space of just days, and sending masks and expert teams abroad to help other countries. Meanwhile, most electoral democracies seemed to fumble early management of the crisis, even if most eventually caught up.

What would a climate technocracy look like? Imagine that, in the face of rapidly rising temperatures and increasingly destructive climate events such as deadly "wet-bulb" heatwaves, catastrophic flooding, or drought-induced famines, the international community creates a Global Climate Council comprising relevant experts from prestigious universities around the world. With enough power, this panel could rapidly implement carbon-abatement policies and override any national legislation that it deems incompatible with its own plans (perhaps with the support of similarly technocratic local climate councils). Though this would likely mean sacrificing millions of jobs or depriving some populations of necessary goods and services, these costs would be justified in the name of saving the planet for the rest of humanity and future generations. What would be wrong about this?

Two big problems stand out. First, while climate change is often conceptualized as one issue, it is in fact many interrelated issues touching on every possible domain of life (agriculture, industry, finance, transportation, energy, education, reproduction, and so on). A niche of independent decision-making that would be strictly related to climate is implausible. The sheer complexity of the issue would introduce all kinds of uncertainties (beyond the intrinsic uncertainty of climate modeling). In the end, climate change would have to recognized as merely one of many issues that bear on humanity's future. In this broader, fuzzier context, it would not be clear that experts have more authority than laypeople.

Second, and more important, climate change is not just a technical issue of prediction, which is why it is a mistake to frame the challenge as a simple question, "How do we achieve carbon neutrality by 2050?" The burdens of widespread decarbonization can be distributed many ways, which means that fighting climate change will always be an eminently political process involving questions of justice and equity. Who should sacrifice what, and for whose benefit?

These are also difficult moral questions. We must make decisions about not just complex chains of physical causality but also about what and whom we should value. While experts will have a lot to say about causal chains, what is likely to happen and why, they have no authority to dictate answers to the second category of questions.

Consider that at the global level, a cap-and-trade policy may seem like a great way to reduce CO_2 emissions and other pollutants but, from the standpoint of justice, it raises obvious questions. Why should the countries that are most affected by climate change despite having a low carbon footprint be given the same restrictions as the countries who have contributed the most to climate change? Why should industrialized countries that are most responsible for climate change be allowed to continue polluting at all?

There are no easy answers to such questions, but one thing is clear: technocrats have neither the legitimacy nor the capacity to make such decisions.

THE JUSTICE IMPERATIVE

Merely acknowledging that climate change ultimately is a matter of justice is not enough. The term "climate justice," critics note, falsely implies that the issue occurs only at the global level and in the aggregate. But for most people, further down the chain of power, climate change is really about local problems like air pollution, toxic water, rising sea levels, and the differential exposure of some communities—typically Black, brown, and Indigenous—to such ills. And, beyond the question of greenhouse gas emissions and rising temperatures, there are also the related issues of biodiversity loss, pollution, and generally declining quality of life.

Moreover, climate justice is not just a matter of distributive justice but also of *corrective* justice. History and geography have placed some countries in the position of being primarily responsible for climate change, and others in the position of being primarily its victims. These historical and geographical patterns also intersect with racial and colonial legacies. Philosopher Olufemi Taiwo thus argues that climate justice should be tied to the question of reparations for slavery and colonization, both between and within countries.

These issues require a more localized lens. Native American journalist Dina Gilio-Whitaker notes that "environmental justice" is a better term, because it focuses the debate on people's immediate surroundings rather than on an abstract entity like climate. Environmental justice requires that we think in more human- and community-centered ways, in terms of the right to healthy, nonpolluted environment. This is what the environmental policy scholar Michael Méndez calls the difficult endeavor of "humanizing climate change."[1]

To address climate change, we need to frame questions in ways that do not separate the technical, causal, and instrumental aspects from values and local environmental concerns. Instead of asking the simple question, "How do we get to net-zero by 2050?" we should ask a more complex one, "How do we get to net-zero, preserve a healthy environment, and conserve biodiversity, while also ensuring that our solutions are fair to all affected communities, and that they account for preexisting patterns of injustice?"

OPEN DEMOCRACY

Given these issues and real-world examples, it should be obvious that the argument for technocracy is in fact an argument for autocracy. The attraction of a technocracy is not that the experts would actually rule best; their advantage over democracy is that they would be able to decide without having to consult the larger population. Their advantage is speed and efficiency, not knowledge per se, much less wisdom.

Consider that even China, the previously mentioned poster child for efficient authoritarianism in the face of a pandemic, recently had to cave in to major popular resistance against its renewed attempts to impose lockdowns to crush a new surge of COVID-19. Since the Chinese population has almost no immunization because of its previous lack of exposure to the virus and the weakness of China's home-grown vaccine, the new variant is proving devastating. Three years after the beginning of the pandemic, the Chinese method no longer seems so successful after all.

Consider also the fate of French president Emmanuel Macron's carbon tax in 2018–2019, which came out of an elected and presumably democratic government but was nonetheless perceived as arbitrarily imposed against people's will. On paper, a carbon tax was a great idea to nudge people away from fossil fuels. In practice, it invited massive political backlash because its effects were unevenly distributed and unfair, punishing exurban working-class commuters while sparing wealthier city dwellers. When the Yellow Vests stormed the streets of Paris, the authorities first resorted to brutal police repression. But in the end, the French government had to take the only acceptable route in a democracy: talking and listening.

Talking and listening took two complementary forms: a "Great National Debate" that lasted two months and involved around two million people, and a citizens' assembly organized at the national level—the Citizens' Convention on Climate. This latter group of 150 French citizens drawn by lot was convened to answer a question that incorporated both the factual nature of climate science and the value judgments of environmental justice: "How to reduce France's greenhouse gas emissions by 40% of their 1990s level by 2030 in *a spirit of social justice*."[1] The mandate of these ordinary French people was to solve a problem that was irreducibly both technical and value-laden. During the 9-month process, the 150 participants insisted on expanding the framework of the question even more, by introducing the question of biodiversity, rights for the environment, and recognition of the crime of ecocide (environmental murder) in the French constitution.[2] On the basis of the resulting 149 proposals, the French parliament then produced its most ambitious climate bill ever, even as the bill fell far short of the Convention's own ambitions.

The lesson here is two-fold. First, no democratic government can reliably force through a policy that the public deems profoundly unfair. The early-stage efficiency of a vertical approach, to climate (or other subjects) policymaking soon gets canceled out by the long process of addressing popular opposition and restoring lost public trust. In fact, even authoritarian governments like China face that constraint of popular resistance to injustice to a degree.

To be sure, a skeptic might say, "Perhaps Emmanuel Macron should have kept using coercion until he broke the resistance. Imagine the impact a fuel tax could have had on French emissions in the two-and-a-half years of deliberation that took place instead, especially with the meager results we know. If the matter is one of life and death, as climate change surely is, we cannot afford to disregard the authoritarian option." My view is that we still have a choice and a bit of time. Given the dismal history, human costs, and high probability of failure of the authoritarian approach, the better bet is to double down, quickly, on an improved version of the democratic deliberative approach.

The lesson of the Great National Debate and the Citizen's Convention, in particular, is not that deliberative democracy takes too long and is ultimately ineffective; it is that such processes and bodies need to be sufficiently empowered by giving them binding rather than purely consultative authority. That is, incidentally, what the French, German, Italian, and UK public think as well, with large majorities supporting the view that governments have an obligation to implement the recommendations of citizens' assemblies.[3] Additionally, the recommendations coming out of such assemblies need to be directly connected to the larger public, for example through nationwide referenda, rather than only to governmental processes, which are often opaque and captured by lobbyists of the fossil fuel industry (although parliaments probably cannot and should not be by-passed entirely).

In my book *Open Democracy: Reinventing Popular Rule for the 21st Century*, I offer an alternative to democratic governance centered around elected officials who engage with their voters only periodically (usually every few years). A better system, I argue, is centered around assemblies of ordinary citizens chosen by lot, who would remain in constant contact with the broader public.

In such a system, democratic feedback loops are revitalized, because politics is no longer primarily the business of professional politicians. Ordinary citizens are re-empowered as sources of law and policy. This model taps into the intuition behind ancient Greek democracy: namely, that each member of the demos is entitled to a say about the good of the polis (of course, membership in the demos would be far more inclusive now than it was two millennia ago).

My vision of open democracy rests on five institutional principles: participation rights, deliberation, majority rule, democratic representation (essentially by lot), and transparency. From a practical standpoint, open democracy might still include a lot of institutions that we are familiar with, such as an elected executive and an appointed Supreme Court. But legislative power would rest in the hands of ordinary citizens rather than elected elites. Like ancient democracy, open democracies would be mostly run by what I call "lottocratic representatives," selected through civic lotteries.

Although there are no current examples of a fully open democracy, we do have ample evidence to show that bodies of randomly selected citizens can and do deliver good recommendations and even quasi-bills (as in France). As an influential and recently updated 2020 Organization for Economic Co-operation and Development report shows, we can draw lessons from close to six hundred cases of sortition-based deliberative processes around the world, many of which have focused on climate issues.

Several recent examples illustrate how citizens' deliberations can lead the way on environmental justice issues. In one example an assembly of twenty-five Icelanders was elected in 2015 from a pool of nonprofessional citizens to build on the work of a randomly selected forum of nine hundred fifty other Icelanders as well as the suggestions and feedback of the larger public, consulted via then novel crowdsourcing techniques, to produce a constitutional proposal that contained landmark provisions to protect the environment. The preamble, for example, made clear that the government should "respect the diversity of the life of the people, the country and *its biosphere*," and that the people aspired "to work towards peace with other nations and *respect for the earth* and all mankind."[4] A key article declared that all natural resources that were not already privately owned would be deemed national property. Further articles granted citizens a right to information about the environment, as well as a right to participate in the preparation of future decisions that may bear on the environment or nature. These were all improvements not only on Iceland's 1944 constitution but also on the set of expert-drafted proposals that the assembly considered as part of its deliberations.

A second, recent example comes from Chile. An elected Constitutional Convention made up of ordinary citizens rather than professional politicians, as per a design choice that overwhelmingly won in a 2020 referendum, produced in May 2022 an even more advanced "green" social contract than Iceland, soon to be put to a national referendum. This constitutional proposal inscribes environmental protection, environmental education, and the right to water, clean air, and a healthy environment as basic human rights.

It recognizes the inherent right of nature to exist and mandates the state to work to prevent, mitigate, and adapt to the climate crisis. It guarantees the preservation of natural commons such as water, air, the atmosphere, and underground resources. It also entrusts an autonomous body to protect the legal rights of nature, building on an Indigenous-led tradition that has been tested at smaller scales in the United States, Ecuador, and India.[5]

A third example, mentioned above, is the French Citizens' Convention on Climate. The 149 proposals put forward by the Convention included making housing retrofitting mandatory under threat of penalties, banning under-two-hour commercial flights where a train alternative is available, promoting vegetarian meals in schools, recognizing a crime of ecocide in the Constitution, and modifying the Constitution's preamble to inscribe the protection of the environment in the state's obligations. One proposal, however, was markedly absent: a carbon tax. Despite being supported by economists and various other experts who advised the convention, the 150 judged it socially too unfair and too controversial in a context where the Yellow Vests movement was still very much alive. In all, the convention did more to address climate justice in 9 months than the French parliament ever did over the last 50 years.

The question now, of course, is whether the rest of the population will endorse proposals made by fellow citizens. In Iceland's case, a referendum on the citizen assembly's constitutional proposal was approved by two-thirds of the voting population (and on the specific question of whether natural resources should be nationalized, 80 percent responded yes). In the French case, convention members decided against submitting most their proposals to a referendum (for reasons that have yet to be fully understood but probably in part because of the plebiscitary and rather undemocratic history of referendums in France). Yet polls show that not only would a vast majority of French people have welcomed the idea of a referendum, they also would have supported all major proposals but one (the reduction of the speed limit on highways from 130 to 110 km/hour). Even the retrofitting mandate, which was both the best lever to reduce CO_2 emissions yet also one of the most coercive and individually costly proposals, met with a surprisingly high 74 percent approval.[6]

In Chile, however, the planned referendum ended up with a rejection of the constitutional proposal, arguably in part because the elected constitutional assembly was insufficiently demographically and ideologically representative of the larger population (skewing too far left). Does this mean a more descriptively representative assembly would not have come up with as strong environmental protections? Perhaps, but it might still have come up with strong enough protections and those would have had more legitimacy and more chances to succeed. This Chilean mismatch is in marked contrast with the almost perfect fit between the recommendations of the French Citizen's Convention on Climate and what the country was seemingly ready to endorse (according to the previously mentioned polls).

The lesson from these examples is that in an era of declining trust people are unlikely to accept solutions to climate change from politicians and experts who seem remote and blind to their experiences. They are more likely to trust people like them—ordinary citizens—to make recommendations about climate justice. But the way these fellow citizens are chosen matters too. Real-world deliberative processes suggest that the public will support the recommendations of randomly selected mini-publics that look and think like they do. In France, polls show that majorities not only approved of the substance of the Convention's proposals but three out of five French people considered the Convention legitimate to fulfill the more general function of making proposals on behalf of the whole population.[7] By contrast, an *elected* assembly, even of nonprofessional politicians like the Chilean Constitutional Convention, is not as likely to be followed by a majority in a referendum, at least if its positioning is perceived as too partisan and far away from the mainstream. The intermediary case of an elected assembly working on the basis of the recommendations of a mini-public is, as the success of the Icelandic referendum suggests, more likely to succeed.

In any case, the respective downstream fates of the Icelandic, Chilean, and French proposals speak to the difficulties ahead. In Iceland, despite the positive outcome of the referendum, the constitutional reform was ultimately blocked by Parliament, which had come under pressure from lobbyists for powerful economic interests who objected to the nationalization of natural resources. In Chile, the constitutional proposal was unambiguously rejected. And in France, only a fraction of the convention's recommendations have been enacted into law or incorporated into regulations, also largely owing to special-interest lobbying. The retrofitting mandate and recognition of the crime of ecocide, in particular, were not passed.

DEMOCRATIZING WORK

As we can see, there is only so much that democratizing the political system can do if powerful corporate interests get in the way and lobby against reforms from within the system.

Democratizing existing political systems will not be enough if we do not also democratize the economy. In other words, we also need to extend our renewed and deepened understanding and practice of democracy to the economic sphere of organizations and industries, especially those that are primarily responsible for the production of greenhouse gas emissions.

Why do we need to do this? First, because we know that the current regime form in most of the economy—shareholder governance or, as Elizabeth Anderson vividly puts it, "private dictatorship"—is not conducive to the transition away from fossil fuels that is urgently needed.[8] Not only it is not conducive to change, as we just saw, it also resists the changes that state regulators, prodded by citizens' conventions, are trying to put in place. By contrast, there are reasons to believe that empowering stakeholders of the firm other than shareholders—crucially, employees—would make environmental and climate concerns more constraining on business decisions. Employees often have qualms and ideas about the environmental costs of firm practices, products, and investments, from air-conditioning temperature and use of single plastics inside firm buildings to the ethics of sourcing materials and dealing with subcontractors to the carbon footprint of various products. It is striking that in most recent cases where business leaders have taken climate change into account in their decisions—in companies like Amazon, Google, or Apple—they have done so in response to pressure from employers, activists, and outside public opinion. Giving these stakeholders more power within firms, at the level of boards of administrators, would only leverage that pressure to much greater effect.

Another reason that democratizing firms and organizations is an environmental imperative is because workers need to have a say in the green transition that will cost them their jobs and livelihood. It is a matter of both dignity and feasibility. Workers who work for Exxon deserve to understand how they contribute to the problem and be given a real choice as to how they can start contributing to fixing it. Additionally, change is likely to happen much faster with than without their adhesion. Even if properly chastised or suddenly enlightened Exxon mobile shareholders asked their representatives on the board of administrators to put the key under the mat, workers and unions would rebel. In other words, in the same way that environmental justice decisions imposed without popular support are likely to trigger Yellow Vests types of social movement, top-down decisions, whether imposed by state regulators or initiated by investors' representatives, will likely be blocked and fought to death by unions and workers, who will feel unjustly punished for environmental sins that are, in truth, those of a much larger number of agents, those of our consumerist, extractive Western model of society as a whole.

Instead of looking at top-down solutions imposed on a hapless workforce, we should learn from the way cooperatives handle crises and transitions. Consider the cooperative Mondragon, a large firm of close to eighty thousand employees that is entirely owned by workers and governed by workers' representatives. During the 2008 crisis, rather than brutally lay off workers, as happened in many of the capitalist firms, the worker representatives at the helm of the company redistributed work and lowered salaries across the board, to keep on as many employees as possible. When it became economically impossible to avoid closing Fagor, one of Mondragon's consumer goods factories, the cooperative provided support to everyone it had to lay off, guaranteeing them jobs in its other companies where possible and continuing to pay employment benefits until every worker who had to leave found new employment elsewhere. It was a humane, respectful way to deal with the crisis, for which everyone in the company took responsibility.

This approach offers a blueprint for a green transition, in which the governance of oil and other fossil fuel companies would work out a humane, respectful way of closing shop while preserving the dignity, hope, and future of workers. Of course, for something as radical as this—more radical indeed than merely surviving a bad business cycle—the states where such international companies are implanted would have to step in, for example with a job guarantee,[9] a basic minimal income, and subsidies and logistical support for training and retooling. Such proposals already exist in the literature, and while it is likely that not all the pain could be made to go away, it is crucial that it be minimized while giving workers in such industries the dignity and social recognition they need to accept and weather that transition.

The workforce in environmentally unsustainable industries is going to resist the necessary shutting down of such industries if the decision is forced on them rather than chosen by them in light of the right state and market incentives. Decarbonization will thus require empowering employees and workers in all kinds of industries. In other words, it will require the democratization of the governance of firms. There are multiple ways to envisage such a democratization, from a version of the German system of co-determination to the cooperative model of worker-owned and directed companies to economic bicameralism, wherein capital and labor each has its own chamber of representatives.[10] Additionally, one may imagine using the selection method of random selection (or stratified random sampling) rather than elections for the choice of worker representatives, as per the arguments made above.

CONCLUSION

Though it may seem counterintuitive, the urgency of climate change demands that we avoid further empowering any particular class of experts. Instead, we must radically democratize our systems at all levels to confer more power on ordinary citizens. That can be done through empowering citizens' assemblies on democratic renovation, enacting constitutional reforms institutionalizing citizen participation and deliberation, and creating new institutions to tackle climate change. At the global level, we can support (financially and otherwise) the next iteration of the Global Assembly on climate, which convened its first pilot session in October 2021 in the margins of COP26. The next version must be larger and more visible. We need to put more pressure on international institutions, and we need to put open democracy on the list of topics for US president Joe Biden's next Summit for Democracy. In parallel, we need to start empowering citizens where a large fraction of them spend most of their time: economic firms and organizations. One first step in that direction is to promote the work and ideas of academics and activist—like the Democratize Work movement[11]—who push for an exploration of novel forms of firm governance and call for moving from destructive hypercapitalism to what might be termed sustainable multi-stakeholder-ism.

Technocracy will not save us from climate change. Empowered citizens in more open forms of democracy still might.

3 CAN DEMOCRACY SAFEGUARD THE RIGHTS OF FUTURE GENERATIONS? CLIMATE CHANGE AND INTERGENERATIONAL INJUSTICE

Daniel Lindvall

It often said that what the founding fathers of American democracy feared the most was the tyranny of the majority. They anticipated that a political power governed by a majority vote could be tempted to restrict liberty in order to remain in power. In his book *Democracy in America*, the French philosopher Alexis de Tocqueville uttered his concerns about the omnipotence of the majority, claiming that the "seeds of tyranny is there if the rights and the ability to everything is granted to whatever power."[1] Unchecked power would thus lead to the demise of democracy. Democracies do not always act prudently, he claimed, since they might "claim to rule upon numbers, not upon rightness or excellence." Alexis de Tocqueville, as well as John Adams and James Madison, argued accordingly for the need to introduce checks and balances among the three branches of power. Moreover, the Bill of Rights was adopted, granting specific rights to individuals and minorities. This was done not only on the basis of moral arguments, but also for the sake of the endurance of democracy.

The constitutional setup, with checks and balances and the Bill of Rights, may have contributed to the endurance of American democracy; however, whether it serves to safeguard democracy for the challenges that it is facing in the near future is questionable. Today, democracies worldwide are challenged by a more alarming and urgent challenge than anything the founding fathers could ever have imagined—but constitutionally, perhaps, not entirely different. We are in a situation wherein a majority of the present electorate has claimed the rights to everything, laying the cost upon the well-being of future generations.

Anthropogenic heating of Earth has already resulted in profound alterations in human and natural systems, and in the near future we can expect more frequent and severe incidents of extreme weather events, sea level rise, and biodiversity loss. Such climate consequences could lead to food insecurity, financial instability, climate-related conflicts, and migration, bringing governing systems and particularly democracy under serious stress. Knowing that greenhouse gas emissions will remain in the atmosphere and affect the planetary system for an unforeseeable amount of time in the future, causing irreparable damage, climate change has an outstanding impact on future generations, constituting an intergenerational injustice. Future generations will need to live with the consequences of the carbon-intensive lifestyle of their ancestors, and in this sense, we are imposing a kind necrocracy on them, restricting their ability to enjoy basic rights and freedom.

In the light of this, we need to ask ourselves whether we can develop and reform our existing democratic system and rearrange the checks and balances among the three branches of power, to make democratic decision-making more farsighted. Is it possible to expand the Bill of Right to protect the unborn against the environmentally destructive short-term decision-making of the current generation? This chapter briefly describes the short-termism of democracy, discusses how democratic decision-making can be made more farsighted and capable of dealing with the climate crisis, and also highlights some of the challenges involved in this exercise.

DEMOCRACY, THE CLIMATE CRISIS, AND INTERGENERATIONAL INJUSTICE

Liberal democracy has often been described as a political system superior to other political systems, in a moral and utilitarian sense. In a triumphal era following the fall of the Berlin wall and the disintegration of the Soviet Union, liberal democracy was praised for the capacity to bring peace, social prosperity, and human development.[2] It was also claimed to be beneficial for the environment and in his book *Earth in the Balance*, Al Gore stated that democracies are "more likely to protect the global environment."[3]

Theoretically speaking, democracies should do more to protect the global environment and to combat climate change. In an open and democratic society, people are better placed to access and spread information on climate change, to organize and form associations, protest, express opinions and concerns, mobilize people in a movement demanding climate action, and, most importantly, to hold unwilling governments to task. Nevertheless, the association between democracy and climate action is highly inconsistent. There are studies demonstrating that democracies are adopting more ambitious emissions control policies, regulations, and targets;[4] however, there is no convincing empirical evidence for a correlation between the level of democracy and carbon dioxide emissions.[5] Several democracies have an utterly poor climate record, particularly the United States, which has actively undermined global climate efforts by, for instance, refusing to sign the Kyoto Protocol and later withdrawing from the Paris Agreement, although it reentered it in 2021.

The reasons why democracies are emitting relatively high levels of carbon dioxide, regardless of ambitious policies, can partly be found in the correlation between economic growth and democratization. Growing income motivates people to expand their rights and freedoms, yet in this process emissions tend to increase. Some empirical evidence demonstrates that democracy can have an alleviating effect on the climate impact of economic growth, yet this is mostly true for high-income countries.[6] However, among developed economies, not a single country is on track to meet the emission targets set in the Paris Agreement.[7] It can be concluded that, regardless of the political system in place, modern societies focus first and foremost on the needs and interests of current generations.

This tendency has been described as presentism or short-termism, and climate policies are for several reasons sensitive to the effects of short-sighted conduct. Psychologists have argued that humans have limited capacity to deal with long-term threats such as climate change.[8] Climate change is evolving gradually, enabling us to psychologically adapt and it is still perceived as something that will affect us in the future or mostly impact people living geographically far away from us. Research shows moreover that individuals tend to devise various barriers to acting, such as externalizing responsibility and blame onto others. They experience a sense of helplessness in the face of a problem that requires collective action at global scale.[9] As long as few people are questioning or changing their carbon-intensive lifestyles, the normative pressure to act is weak.

These psychological and social barriers to climate action can obviously stymie the agency of a political system that rests on the will of the people. In the midst of the climate crisis, politicians are elected and reelected despite their failure to act. Although people have become more concerned in recent years, climate action is still not on the top ten priority list of voters in the United States, ranking below issues such as dealing with terrorism and immigration.[10] Climate concerns will probably increase as more people experience extreme weather events, and this could have an effect on political behavior. Research on European voters has for instance demonstrated that experience of climate extremes, heat episodes, and dry spells can influence people's political decisions.[11]

It can nevertheless be argued that existing liberal democratic institutions are not well designed to deal with dilemmas that are not perceived to be acute and that require global collective action. Democratically elected politicians are first credited for their accomplishments within the nation-state, and not those in the international arena. The time-remit of democratic decision-making is also limited by the election cycle. Politicians who implement the measures that climate science demands may not be reelected, since their decisions are not necessarily popular with the electorate nor do they result in any instant emission reductions. Carbon taxation is an example of a measure recommended by several climate policy experts that could lead to unpopular short-term consequences, such as increased energy costs, while its impact on emissions will not materialize within the time frame of a governing mandate. Surveys in both Europe and the United States show that people in general are not overly enthusiastic about carbon taxes,[12] and in France an attempt to introduce such a tax resulted in the protest movement known as the Yellow Vests movement (*mouvement des gilets jaunes*).

In most developed nations, there is also a short-term bias in democratic decision-making, as young voters, who have the greatest interest in a sound future, are in a minority position in the electorate. In the 2020 US presidential election, millennials made up less than one-tenth of registered voters, and the turnout of this constituency was only around 50 percent compared to 75 percent of voters above age sixty-five.[13] In effect, the incentives for politicians to prioritize issues relevant to future generations, such as climate change, are weak. Citizens who may live to experience the more system-threatening effects of climate change are still not eligible to vote and will not be able to hold current elected leaders accountable.

Furthermore, the costs and benefits of various climate mitigation measures might be uncertain, as different technological and scientific innovations are speedily and simultaneously developed. Even though new innovations can replace ineffective and carbon-intensive technologies, there is always a risk of lock-ins when governments choose to invest in one specific technological system. Global warming is also an issue characterized by complexity and interconnectivity, and decarbonization of individual economic sectors might have multiple and often unpredictable economic, social, and cultural consequences. Voters might therefore punish governments for the negative side effects that policy-makers failed to foresee. On the other hand, this interconnectivity may generate co-benefits and climate action could, for instance, lead to economic, health, and security benefits, such as improved air quality in urban centers or energy independence.[14] Focusing on positive side effects and ensuring that mitigation measures generate certain short-term paybacks is obviously essential for climate policy success. Surveys show, for instance, that people tend to accept carbon taxes if the revenues are redistributed back to ordinary people,[15] and they are generally positive toward renewable energy, if such investments lead to lower energy prices, create jobs, and generate profits for the communities directly affected by them.[16]

On the other hand, the complexity and uncertainty of the climate transition can be exploited by interest groups, such as fossil fuel companies, spreading disinformation and thereby delaying climate action. The fossil fuel industry is a powerful economic entity, focused on short-term economic gains, and has evidently influenced the climate policy output of individual countries. This is the case in the United States, where the fossil fuel industry, and a number of think tanks, have used their economic muscle to capture climate policies and outmaneuver the climate movement. According to research by sociologist Robert Brulle, the lobbying expenditure of the fossil fuel industry relating just to climate change legislation in the US Congress was, from 2000 to 2016, approximately ten times larger than the total resources spent on lobbying by the environmental movement and the renewable energy industry.[17]

To successfully safeguard future generations, democracies must be able to withstand the pressure of lobbying and influential interest groups. Private financing of political campaigns needs to be properly regulated, and political parties should preferably be publicly financed. This is essentially a matter of capable, strong, and independent state institutions, working with a low level of corruption. In a study of the associations among the level of democracy, corruption, and carbon dioxide emissions, the political scientist Marina Povitkina found that democratic governance reduces emissions levels, but only in countries with low levels of corruption.[18]

TOWARD MORE FARSIGHTED DEMOCRATIC DECISION-MAKING

In recent years, several governments have begun to pursue more active and stringent climate policies, and this has had an effect on emissions. A total number of 195 countries and the EU have signed the Paris Agreement, with the goal of keeping warming well below two degrees Celsius compared to preindustrial levels, and an ambition of keeping it below one and a half degrees. The agreement requires countries to submit their national determined contributions, thereby motivating them to develop their climate governance capacities and adopt national policies, legislation, and emissions-reduction targets. According to the Grantham Research Institute of the London School of Economics, more than 2,100 different forms of climate-related legislation or other policy instruments had been adopted worldwide up to 2022.[19]

Although climate acts and emissions-reduction targets are generally not binding instruments with sanctioning capacity, they can help to overcome short-termism and influence the development and implementation of climate policies. Studies show that, in countries with a strong rule of law, where legal provisions are likely to be followed, climate acts and a legal framework for assessing climate policies can help to bring down emissions.[20] Moreover, an increasing number of countries have introduced some kind of pricing mechanism on carbon emissions, such as emissions trading systems (ETS) or carbon taxes. In 2022, sixty-eight carbon pricing instruments had been implemented worldwide, covering around 22 percent of global greenhouse gas emissions.[21] Setting a price on carbon can be considered as a method for integrating long-term climate goals into the economic system, and it has shown to be an effective method for curbing emissions.[22]

The development of climate-related legislation, targets, and pricing mechanisms has moreover led to an institutionalization of climate policies. Climate action is shifting from being just a political subject to becoming an issue for bureaucratic oversight and juridical review. This tendency is ultimately expressed by constitutionalizing environmental care and protection, a practice that has been ongoing in recent decades. According to one study, 148 out of 196 constitutions worldwide contain provisions referring to environmental protection.[23] At least eleven constitutions have specific climate clauses,[24] and a few are literally referring to the interest and needs of future or coming generations, generally in relation to the protection of the environment. The German constitution, for instance, explicitly express that the state has "responsibility towards future generations," and "shall protect the natural foundations of life and animals by legislation." Some countries have even entitled nature with specific rights, as in Peru and Ecuador.

These constitutional principles are, however, often formulated as broad and visionary principles, and cannot be considered as legal obligations or actionable duties. Nevertheless, as discussed below, with the pressing nature of the climate crisis and increasing number of climate litigations, the legal interpretation of climate acts and constitutional principles could be altered, expanding the rights of future generations, and thereby also affecting the boundaries of current democratic decision-making.

PROXY REPRESENTATION OF FUTURE GENERATIONS

In several ethnically heterogeneous societies, minority groups are entitled with specific rights within the democratic system, such as reserved seats in a parliament or communal representation, ensuring the inclusion of their voices and interests in decision-making processes. In the discussion on intergenerational injustice, it has been suggested that the interest and needs of future generations could be expressed by a similar kind of proxy representation. This could possibly be done by allocating seats in parliament to representatives of future generations or by creating institutions commissioned to act in the interest of the unborn.[25] Such institutions have been set up, for instance, in Wales, Scotland, Canada, Israel, and Hungary, although the effect of these initiatives has so far been limited.

A more concrete measure in this regard is to lower the voter age to sixteen years, as has been done in Austria, Brazil, Ecuador, Argentina, Scotland, and Wales. An evaluation of the electoral outcomes in Austria shows, however, that young people might vote not only for parties with progressive environmental agendas, but also for right-wing populist parties, which are often opposing climate action.[26]

When it comes to the challenges of climate change, the interest of future generations might be best represented by scientists, alerting policy-makers on the need of action. Scientific experts have traditionally been involved in all kinds of policy areas; however, in the case of climate change, science has been at the very center of policy-making processes. At the global level, climate policy is guided by the UN Intergovernmental Panel on Climate Change (IPCC), which had a key role in formulating the UNFCCC and the Paris Agreement. An increasing number of countries have furthermore established national climate policy councils, which are tasked to provide science-based advice and assessments of governmental policies. These are often independent bodies, whose assessments and recommendations are impartial, although in some countries they are part of the government structure. They can be established as a requirement of climate acts, and normally consist of scientific experts, while in some cases also including representatives of public authorities, civil society, or other stakeholders.[27]

These expert bodies can contribute to evidence-based decision-making, continuity, and long-term goal achievement, and thus generate action on climate policy. They can also be beneficial for the democratic decision-making process, by identifying the success and failure of pursued policies and thus enlightening the electorate and strengthening accountability mechanisms. In this sense they can be seen as representatives of future generations, and as a positive response to the political slogan "listen to the scientists," often uttered by the youth-dominated climate movement.

Involving scientific experts in decision-making can be problematic, however. The role of scientists in policy-making is often arbitrary, and politicians can choose to dismiss or heed their advice as it suits their political agenda. Climate policy councils can help to formalize the role of scientists in the policy-making process, but recommendations are often very general and do not refer to specific policy measures. On the other hand, as a result of the institutionalization of scientific experts in climate policy-making, scientists are becoming increasingly integrated into governing bodies, and thus endowed with a certain degree of political authority. The rulings in several climate-related court cases has also been based on scientific advice.

With the growing relevance of climate policies, more frequent cases of climate litigation, and the potential binding character of climate targets, science can unwittingly be turned into a new fourth branch of power. This tendency could result in a problematic politicization of science and constitute a challenge to the independence and integrity of scientists. Allowing public authorities, such as environmental protection agencies, to interpret and enforce policies to keep emissions in line with set targets and legislation could also lead to an undemocratic inclination to technocratic governance increasing the power of the executive branch over the legislative. It is necessary to listen to science in order to fully understand the problems and risks at hand, but democratically elected officials are generally better placed than are scientists for dealing with the conflicting aims and interests involved in the climate transition process. Developing appropriate models for science-policy interaction in democracies, and crafting the right balance between the competencies of appointed experts and elected officials, will be of great importance.

FIGHTING FOR THE RIGHTS OF FUTURE GENERATIONS IN THE COURTS

An increasingly popular form of climate action is legal activism, encouraging courts to coerce governments to adopt more ambitious climate policies. Recent years have seen a surge in climate-related court cases. Between 1986 and 2015, around eight hundred climate-related legal proceedings were documented worldwide; in the six following years, more than a thousand climate litigations took place.[28] At least thirty-seven cases have been brought directly against governments for delaying the climate transition, for failing to act in accordance with set goals, for violating climate-related legislation, and for authorizing environmentally degenerating activities. Business enterprises, and especially fossil fuel companies, have also been accused of misleading the public or of exposing people and nature to severe climate consequences through their emissions.

This climate-related legal activism is partly a result of the frustration expressed by environmental organizations and individuals toward inadequate climate policy measures. Individual successes, in particular the so-called Urgenda case in the Netherlands, have motivated activists to turn to the courts. Moreover, the adoption of national climate legislation and targets, and the ratification of the Paris Agreement, constitute a legal foundation for the litigation processes.[29]

So far, the outcomes of climate legal activism have been mixed. There are several difficulties involved in the processes, such as dealing with scientific evidence and establishing an association between the suffering of the plaintiff and the actions of individual companies or states, as well as the conservative approach of the judiciary. The climate cases differ depending on the legal systems of different countries and how the litigation processes have been conducted; and therefore it can be difficult to make general assessments of the effectiveness of climate litigations.

Nevertheless, a few court cases have been successful. One of the most significant is the aforementioned Urgenda case, initiated by Dutch citizens who accused the state of the Netherlands of acting contrary to the constitution and, above all, of violating Articles 2 (right to life) and 8 (right to respect for private and family life) of the European Convention on Human Rights (ECHR) by its failure to reduce greenhouse gas emissions. The case was decided in a lower court in 2015, which required the Dutch state to reduce emissions by 25 percent by 2020 (in terms of 1990 levels), and the ruling was upheld by the Supreme Court in 2019.[30]

Other successful climate-related cases include the Leghari case in Pakistan in 2015 and court cases against the Irish government in 2020 and the French and German governments in 2021. In May 2021, a Dutch court required the energy group Royal Dutch Shell to reduce its emissions by 45 percent by 2030. This case is the first major climate lawsuit successfully pursued against a private company, and several similar cases against fossil fuel companies are ongoing in France and the United States.

CAN CLIMATE LITIGATION HELP EXPAND THE RIGHTS FOR FUTURE GENERATIONS?

It is difficult to judge what influence legal activism has had and could potentially have on climate polices. Some of these court decisions, such as the Urgenda case, are not requiring governments to undertake any specific measures, but simply calling for general emission reductions. It is unlikely that governments that fail to comply with court decisions will face any legal consequences. Legal processes are also very costly and slow, and the outcomes often uncertain. Nevertheless, court decisions can be seen as powerful tools for applying political pressure, and can be used to raise awareness and mobilize people for policy demands. In the "Sixth Assessment Report of the IPCC on Mitigation of Climate Change," the panel recognized the trend of climate litigations as a method that can "affect the stringency and ambitiousness of climate governance."[31]

Cases that are brought against business enterprises might be particularly effective. Fossil fuel companies are obviously responsible for releasing massive amounts of carbon dioxide in the atmosphere, but some of them have also been involved in active disinformation campaigns and have coerced governments to disregard scientific advice. Climate litigation directed to such corporations could influence decisions of investors and insurance companies, inspire divestment activities, and potentially prevent hydrocarbon exploration projects.

Most important, climate litigation processes can contribute to the development of jurisprudence, generating new legal obligations and entitling citizens with new rights. What several of these court decisions show is that the climate issue is a matter of justice, and that certain constitutional principles related to civil rights and freedoms are important in this context. A court case that clearly demonstrated this was the decision of the German Constitutional Court in April 2021, *Neubauer et al. v. Germany*. According to the decision of the court, the German Climate Protection Act and the mitigation measures of the government were considered to be contrary to fundamental rights because they leave too little room to reduce emissions after 2030, thus insufficiently ensuring the rights and freedoms of future generations. The Neubauer case demonstrated that the wording of the German Constitution regarding the "responsibility for future generations" may have legal implications, as well as engender international commitments, such as the Paris Agreement.

Moreover, a few climate-related cases are currently pending before the European Court of Human Rights in Strasbourg, and it is not unlikely that the court could rule in favor of the plaintiff in at least one of these cases, recognizing that inadequate climate policy implementation can be seen as a violation of human rights.[32] Successful climate-related litigation can thus contribute to change and expand the institutional framework of democracy, demanding that political decision-making take greater account of the rights and interests of future generations.

From a democracy perspective, however, it is not entirely clear whether climate litigations directed to individual states will have only positive effects. When the climate issue is taken to court, the dynamics of democracy can be changed, shifting the climate-policy process from the parliamentary arena to the judicial system. This could possibly demotivate individuals from participating in the ordinary processes of democracy. Moreover, when judges are asked to overrun legislative processes and oppose decisions endorsed by a majority of the electorate, they may violate the separation of powers and undermine the mechanisms of accountability. An example of a case dismissed partly on the basis of the separation of power argument is *Juliana v. US Government*, in which the government was accused of violating individuals' rights to life and liberty by urging and allowing the burning of fossil fuels.

In a constitutional democracy, the judiciary may challenge democratic decisions, but only if its decisions are judged to be in violation of the constitution and fundamental rights, and thus endangering the endurance of democracy. Obviously, ignorance of the climate crisis can be seen as a threat to fundamental rights; however, climate litigation, such as the Urgenda case, is still controversial, since it can disqualify the formal procedures of democratic decision-making and thus challenge the legitimacy of the democratic system.[33] Court litigations against individual destructive policy measures, such as decisions to undercut emission regulations or issuing oil-drilling permits, might be unproblematic, but asking courts to assess climate policies broadly can be politically divisive, knowing that climate change as such is a political issue that permeates all spheres of society. Moreover, climate litigation may not only politicize the courts, but could also, as discussed, bring scientists into the position of judges.

Court litigations are, on the other hand, a double-edged sword that could also be used by interest groups representing the fossil fuel industry, asking courts to overrule or block mitigation measures issued by public authorities. This was the case in the United States, where the authority of the Environmental Protection Agency (EPA) has been challenged in court proceedings, such as *West Virginia v. EPA*. In this case, the Supreme Court, controlled by conservative-leaning judges, restricted the authority of the EPA to regulate certain emissions. This case demonstrates that, rather than empowering the judiciary, climate litigation could be used to undermine climate policy stringency, challenge the power of the executive branch to interpret and enforce adopted legislation, and disturb the balance of the three branches of power. In conclusion, it is unclear whether judges overall will act in the interests of future generations or defend short-termism and a business-as-usual approach.

CONCLUSION

In the coming years, more people will experience the direct consequences of global warming; hopefully this will lead to an increasing demand for political action. Knowing that the level of carbon dioxide in the atmosphere is already alarmingly high and is beginning to seriously destabilize natural systems, those in charge of climate policy-making will be focused not only on emissions reductions, but also on adaptation and crisis management. Climate policy will be approached as an issue requiring instant response, yet this might also generate greater crisis awareness and stimulate more farsighted approaches. Climate policies will furthermore be increasingly institutionalized, as regulations, national and international carbon-pricing mechanisms, and policy institutions such as advisory councils, are developed. At the same time, a growing number of individuals affected by various extreme weather events will become eager to seek justice. These developments will obviously affect climate policy-making, and possibly also the framework of democratic decision-making.

The pace of these trends will probably differ in different countries, depending on legal traditions, political systems, fossil fuel dependency, climate policy development, and the strength of social mobilization. In countries such as the United States, where climate policies are deeply divided and polarized, climate-skeptical political forces will continue to fight back, trying to undermine policies, regulations, and institutions and influence the court system to act in the interest of the fossil fuel industry. Fighting for climate action and the development of democracy will for sure not be a peaceful process, but will undoubtedly be associated with both legal and political conflicts.

Climate litigation could possibly be a weapon in this struggle, contributing to stronger regulations and policies; however, the court system could also be used against climate action, blocking policy decisions. It can be concluded that the social transformation that ultimately may safeguard future generations must be realized through political engagement and mobilization of people demanding a sound future. In recent years, the climate movement has grown stronger, and to a certain extent, intergenerational injustice has been a mobilizing and empowering factor. By embodying and acting as representatives of future generations, youth-dominated movements such as Fridays for Future and the Sunrise Movement have managed to communicate a concrete expression of the immorality of climate change. Along with this development, greater segments of the electorate will hopefully be persuaded of the severity of the climate crisis and demand political action. The strongest transformative forces of democracy are found, at the end of the day, in the collective power of people mobilizing for a morally convincing cause.

II ROADBLOCKS

4 OUR URGENCY OF NOW: CONVERGING GLOBAL CRISES IN A TIME OF POLITICAL EVOLUTION

William J. Barber III, JD

America has defined herself as we the people, a nation whose noble statement of identity says, "We know these truths to be self-evident, that all persons are created equal, endowed by their Creator with certain inalienable Rights, that among these are Life, Liberty and the pursuit of Happiness." America has said that the establishment of justice is the first principle of our nation, and yet Dr. King once critiqued this nation he loved so much, saying, "Let us be dissatisfied until America will no longer have a high blood pressure of creeds and an anemia of deeds." "America, oh America, it's never been America to me," Langston Hughes once said.[1] "But I swear this oath America will be."

America—yes there has been progress, but it has come only when individuals who knew who they were and who knew who America should be dared to challenge this country and not give up on her possibilities.

Bishop William J. Barber II

* * *

On August 28, 1963, some 250,000 Americans gathered in the nation's capital to take part in the March on Washington for Jobs and Freedom. At this march, Rev. Dr. Martin Luther King took the podium, where he first articulated the idea of a fierce urgency of now—an idea that there was such a thing as being too late. He urged the nation to recognize that it was not a time for apathy or complacency but, instead, it was a time for "vigorous and positive action."

On April 4, 1967—exactly one year prior to his assassination in 1968—King once again uttered the refrain at the esteemed Riverside Church in New York City. There he addressed a diverse group of more than three thousand for the pacifist organization Clergy and Laity Concerned about Vietnam. Illustrating his shift in focus to economic justice for all Americans and challenging the nation on the triple ills of poverty, militarism, and racism, he was accompanied by Amherst College professor Henry Commager, Union Theological Seminary president John Bennett, and Rabbi Abraham Joshua Heschel. There, King declared,

> We are now faced with the fact that tomorrow is today. We are confronted with the fierce urgency of now. In this unfolding conundrum of life and history there is such a thing as being too late. . . . Over the bleached bones and jumbled residue of numerous civilizations are written the pathetic words: "Too late." There is an invisible book of life that faithfully records our vigilance or our neglect. "The moving finger writes, and having writ moves on." . . . We must move past indecision to action. . . . We must find new ways to speak for peace . . . and justice throughout the developing world—a world that borders on our doors. If we do not act we shall surely be dragged down the long dark and shameful corridors of time reserved for those who possess power without compassion, might without morality, and strength without sight. Now let us begin. Now let us rededicate ourselves to the long and bitter—but beautiful—struggle for a new world.[2]

Dr. King's words ring true with renewed vigor as we face our own fierce urgency of now in the form of multiple societal upheavals of unprecedented scale. The three crises of poverty, democracy, and climate form a nexus in which each one of us living must reevaluate and redefine our existence. All of us stand here together at an extraordinary—and perilously dangerous—moment in time. For the planet, for our countries, for the people of the world. Especially for the most vulnerable and most marginalized among us.

ON POVERTY

> An imbalance between rich and poor is the oldest and most fatal ailment of all republics.
>
> —PLUTARCH, GREEK HISTORIAN

Regarding the crisis of poverty—in a nation that has acquired the greatest amount of economic wealth presently on the planet and in the history of mankind—we still see today more than forty million American citizens subsisting below the federal poverty line. That number jumps to more than ninety-five million when using the US Census Bureau's official poverty measure and jumps again to more than 140 million—nearly half of the US population and including fourteen million children—when using the supplemental poverty measure.[3] We see, prior to the impacts of the global pandemic of COVID-19, more than seven hundred American citizens dying per day because of the brutal effects of poverty. We see, in the wake of the brutality of COVID-19, growing health-care costs resulting in tens of millions of people not able to get the treatments they need; eighty-seven million people who are uninsured or underinsured; 16.2 million people who have lost their employer-provided insurance; rising hunger costs resulting in food insecurity for more than forty-one million Americans; and $160 billion per year in health-care costs for poor families; nearly one in three Americans are at risk of not being able to afford water in five years.[4]

King spoke prophetically to us on this issue in 1967: "There are forty million poor people here. And one day we must ask the question, 'why are there forty million poor people in America?' And when you begin to ask that question, you are raising questions about the economic system, about a broader distribution of wealth."[5]

The issue of poverty has often been the subject of political performatives—either heavily weaponized as an issue of personal work ethic or racialized as an issue concerning only Black and brown people. Both characterizations are incorrect, with the vast majority of poor Americans—nearly sixty-two million people—existing as "working poor" individuals with incredible work ethic, often working multiple jobs and long hours and yet still unable to make ends meet. The racialization of poverty is equally deceptive. Although Black and brown people have higher rates of their populations by percentage impoverished, white Americans continue to make up the majority of poor people in the country based on raw numbers—approximately sixty million people.

The disproportionate rates of poverty for communities of color can often be traced to the effects of intentionally discriminatory government policies designed to deny economic opportunity for these communities. Policies such as rescinding promises of land grants to formerly enslaved individuals, the practice of redlining, forced displacements through eminent domain, predatory lending, and racial pay gaps have kneecapped economic opportunity for these communities over the course of generations.

Adding insult to injury, many of the same political voices that assign blame to fellow American citizens—proposing that it is their fault for being poor, instead of actually solving the issue of national poverty—often position themselves as the most ardent voices for lofty tax breaks for corporations and the extremely wealthy. Without fail, we often see in these voices a vicious sampling of the failed theory of "trickle-down economics," the dog-whistles of the Southern Strategy, and the phenomena articulated by President Lyndon B. Johnson: "If you can convince the lowest white man he's better than the best colored man, he won't notice you're picking his pocket. Hell, give him somebody to look down on, and he'll empty his pockets for you."[6]

Instead of being taken for the national security crisis it is, this narrative proves dangerously effective in undermining populist campaigns across lines of race and class, and instead creates national stagnation on solving the actual issue.

ON DEMOCRACY

> It is for us, the living, rather, to be dedicated here to the unfinished work, . . . that this nation, under God, shall have a new birth of freedom—and that, government of the people, by the people, for the people, shall not perish from the earth.
>
> —ABRAHAM LINCOLN, SIXTEENTH PRESIDENT OF THE UNITED STATES OF AMERICA

Regarding the crisis of democracy, it must first be recognized that the sole avenue to policy action on every issue of social importance is through free and fair elections, in which the American people are able to choose candidates responsive to our interests, issues, and demands. From the eradication of poverty to bold action on climate, and from the protection of women's rights to strengthening public education to expanding and protecting voting rights themselves—the path to achieving systemic policy change goes directly through the ballot box.

Over the last decade, our nation has witnessed coordinated attacks on voting rights, hyperpartisan attempts to gerrymander electoral districts, attempts to hijack our nation's judiciary, and efforts to infuse electoral cycles with unlimited dark money. Taken singularly, any one of these occurrences would be alarming. The reality of all four indicates the necessary conditions for a twenty-first-century democratic crisis.

The attack on voting rights is especially sinister. Along with access to quality education, the ability to exercise one's right to vote is one of two critically fundamental pillars through which an individual can operate in their essential role as a functioning citizen. Access to education allows an individual to shape their views and opinions of the world, comprehend their present reality, and define their needs, wants, and interests. The ability to exercise their right to vote allows that comprehension to be heard and translated into policy. Education allows one to conceptualize what society should look like. Access to the vote facilitates societal shaping.

The historic synergism among access to the vote, access to education, and full participation as a citizen is colorfully illustrated in the postwar era of Reconstruction in the South. After the passage of the Reconstruction Amendments, formerly enslaved African Americans began to take their place as voters and legislators in Southern states. To secure the right of equal citizenship as established in the Fourteenth Amendment, African Americans demanded the establishment of state-funded public school systems. African American sociologist W. E. B DuBois outlines this push in *Dubois on Education*: "The first great mass movement for public education at the expense of the state in the South came from Negroes. . . . Public education for all at public expense was, in the South, a Negro idea."[7]

DuBois continues, in *On Education*, describing how as African American voting increased, citizens were more successful at petitioning Congress in the distribution and administration of federal funds to support public school systems. This in turn enraged many of the wealthy Southern white elite, who would eventually lead efforts to disenfranchise Black voters en masse, segregate school systems, and erase years of progress on public education (for both Black and pauper white children) in Southern states.

This history illustrates the checkered past of voting rights in America. Where voting rights have been expanded, they have often led to elevated status for marginalized communities and widespread societal progress for the nation. Where voting rights are narrowed, they have often led to less representation, historic rollbacks of societal progression, and irreparable harm. This history also puts into perspective why broad access to the ballot is critical to a functioning democracy: it is a matter of power: power to shape the society in which one lives; power to have one's voice and opinions heard; power to ensure accountability of the nation to the citizens themselves. And power itself is a gateway to freedom.

On July 4, 2019, esteemed historian Ibram X. Kendi published an article (adapting Frederick Douglass' July 4, 1852 address) titled "What to an American Is the Fourth of July?" In it, Kendi writes,

> Power comes before freedom, not the other way around. Power creates freedom, not the other way around. We can't be free unless we have power. Freedom is not the power to make choices. Freedom is the power to create choices. And to have the power to shape policy is the power to create choices. That is why power is in the hands of the policymaker. . . . America is the story of powerful people struggling to keep their disproportionate amount of power from people who are struggling for the power to be free.[8]

The story of voting rights in our country is bittersweet. Since the early days of the American experiment, the question of who gets access to the vote—and thereby to power—has been contentiously debated. In the earliest days of America's great democratic experiment, the right to vote was narrowly granted to the class of white male landowners. In certain cases, states also employed religious tests to ensure that only Christians who were also a part of this subset would be allowed access to the ballot.

In the beginning of the nineteenth century, states began to relax the property requirement for voting. Later, in the middle of the century, and post-Civil War, the Reconstruction period saw the passage of the Fifteenth Amendment, guaranteeing the right to vote to Black *men* (with the glaring absence of participation for both Black and white women). The amendment was fully ratified by the states in 1870, and the period afterward saw a radical increase in Black voter turnout, as well as representation in governance—with more than 1,500 elected Black people serving in office, including, two US senators, Hiram Rhodes Revels and Blanche Bruce, both of Mississippi. This period of Reconstruction and radical representation was eventually dismantled, succumbing to deeply entrenched resistance to Black enfranchisement via the implementation of barriers such as literacy tests, poll taxes, the "grandfather clause," and outright racial violence, designed to intentionally dissuade Black men from exercising the vote. With the passage of the Nineteenth Amendment to the constitution in 1920, American women saw decades of advocacy turn into victory, securing access to the ballot.

In the 1960s, the nation's attention was once again turned to the South, via the focus of the civil rights movement on the racist system of disenfranchisement for Black voters. In 1965, Rev. Dr. Martin King and allies launched a campaign of civil disobedience to highlight the brutal and systemic suppression of Black political power in the South. This came on the heels of the success of challenging the Southern policy of segregation in the decade before. On Sunday, March 7, 1965, in an event known as Bloody Sunday, the brutality of Southern racists in enforcing Black disenfranchisement was put on full display, shocking the nation and spurring congressional action for renewed voting rights legislation. Less than two weeks after the event, the Voting Rights Act was introduced in the House and Senate, and the act passed on August 4 of that same year. Despite the act's passage, federal troops were still necessary to ensure compliance for several years. The following year, in 1966, in the case *Harper v. Virginia State Board of Elections*, poll taxes were found to be unconstitutional.

This history provides context for presenting attempted rollbacks on voting rights as the crisis they represent. Specifically, understanding how federal legislative power and physical power were necessary in 1965 to guarantee the freedom promised to Black Americans over a century ago gives Kendi's statement enhanced meaning: that "power comes before freedom, not the other way around. Power creates freedom, not the other way around."

Recent actions by the federal government suggest devastating amnesia in both the fight for voting rights in this country and in the government's necessary role as a powerful counterbalance to ensure political power remains in the hands of the American people—all of the American people.

The election of 2008 was a glimpse into what political power in the hands of the American people could look like. More significant than the election of a particular candidate was that the electorate was the most racially and ethnically diverse in US history, with nearly one in four votes cast by nonwhites, an unprecedented number at the time. The nation's three biggest minority groups—Blacks, Hispanics, and Asians—each accounted for an unprecedented share of the presidential vote in 2008. Unprecedented, but not unexpected: as America goes through projected racial demographic shifts, absent systemic barriers to access, more people of color will be eligible and active in the electoral cycles. The year 2008 ushered in the most racially diverse electorate in US history; in 2012, the electorate was again the most racially diverse, and the rate of Black turnout exceeded white turnout percentage-wise for the first time in history; in 2016, the electorate was once again the most racially diverse electorate in history, but also saw the winning presidential candidate, Donald Trump, win the Electoral College vote while losing the popular vote by more than 2.8 million votes.

This occurrence—besides being only the fifth time in our nation's history of happening—serves as a modern case study on the vestiges of targeted attempts by some to mitigate the full enfranchisement of the masses. The Electoral College, for example, was originally proposed by the wealthy planter James Madison in conjunction with the Three-Fifths Compromise—an instrument designed to bloat the power of Southern states in presidential elections by allowing them to include enslaved populations, at a rate of three-fifths per enslaved individual, in the overall state population tally to determine their state's apportionment of electoral votes.

This effectively meant that enslaved individuals would effectively boost the political power of the people who had enslaved them for economic gain. This allowed Southern states—specifically, wealth planters in those states—to wield considerable influence in the presidential electoral cycle, a phenomenon still very much felt in modern politics. In a similar fashion, the promise of a return of power to a privileged few was the impetus behind the Compromise of 1877, through which presidential candidate Rutherford B. Hayes agreed to end federal enforcement of Reconstruction and cede power back to Southern states in exchange for their delivery of the contested presidency, leading to an era of white supremacy and Jim Crow. This promise of a return of power was the impetus behind Harvey "Lee" Atwater's Southern Strategy, through which the Republican Party embraced securing political support among whites by playing into the fracture point of Southern racism and dog-whistle politics in the wake of the gains of the civil rights movement.

And this promise of a return of power was unleashed recently, once again and more broadly, in numerous forms including the renewed launch of Republican-sponsored state voter suppression efforts following the 2008 election and the historic turnout of people of color; the rise of groups such as the Tea Party and Birther movements, which effectively utilized racial and polarizing dog-whistles to appeal to their base; the 2010 decision in *Citizens United v. FEC*, wherein the court ruled by a five-to-four decision that corporations should essentially be considered the same as people with the right to spend unlimited sums of money in political campaigns, boosting the political power of wealthy elites and ushering in an unprecedented wave of anonymous campaign finance, allowing dark money to flood our nation's political discourse.

The promise of a return to power was further unleashed in 2012, with another wave of state voter suppression efforts, resulting in more than seventeen states implementing a variety of voter suppression tactics, including hurdles to registration, reduction of early voting, and strict voter identification requirements. It was further unleashed in 2013 when, in another five-to-four decision, the Supreme Court's conservative majority in *Shelby v. Holder* ruled that the coverage formula of Section 4b of the Voting Rights Act was unconstitutional and outdated, resulting in, essentially, the gutting of the entire act. Ironically enough, the court looked to the incremental gains in voter turnout of people of color as indicators that voter discrimination was no longer an issue and the act was no longer needed, rather than as a confirmation that the design of the act was just starting to function, nearly fifty years later. Since the ruling in *Shelby*, twenty-seven states have passed discriminatory voter suppression laws targeting people of color and student voters, and 1,700 polling places have been systemically closed, limiting voter access.

We see this long list of recent democratic crises come to a head in 2016, when, in the first presidential election in fifty years without the full protection of the Voting Rights Act, Donald Trump is elected to the US presidency despite having lost the popular vote by 2.8 million votes. And, despite the theatrics of the Trump administration, the fragility of our democracy is clearly made evident on January 6, 2021, when a violent mob of insurrectionists, fed by years of propaganda and rhetoric, attempted to storm the US Capitol and prevent the formalizing of the victory of the duly-elected new president.

This long history again reiterates that democracy is a question of power. Whether that power is well-distributed among the masses or held for a select few is to be determined. It is a question we must quickly answer, for in the answer lies the gateway to our freedom, and our future.

ON CLIMATE

> If we do not save the environment, then whatever we do in civil rights will be of no meaning, because then we will have the equality of extinction.
>
> —JAMES FARMER JR., COFOUNDER OF THE CONGRESS ON RACIAL EQUALITY

In 2019, The UN Special Rapporteur on Extreme Poverty and Human Rights, Philip Alston, issued a report warning, "Climate change threatens to undo the last 50 years of progress in development, global health, and poverty reduction. . . . It could push more than 120 million more people into poverty by 2030 and will have the most severe impact in poor countries, regions, and the places poor people live and work."[9]

In 2020, at an address to the European Economic and Social Committee, UN Climate Change deputy executive secretary Ovais Sarmad called 2020 "a critical year for addressing climate change." He further stated that "climate change [had] morphed from a serious challenge to a full-blown emergency—one affecting almost every part of the world. . . . Climate change does not respect politics, but our politics must respect climate change."[10]

Climate change is the existential crisis of our time. A product of our more than century-long dependence on fossil fuels to power our economy, the crisis looms as a force multiplier for every form of historic inequity we as a global society have tolerated—and perfected—over the course of time. At its origins, the need to power our economy was lauded as a necessity for economic growth and human progress. However, its buildout came at the expense of vulnerable communities that were deemed disposable for the greater good—which is why, both globally and here in the United States, the buildout of climate-changing fossil fuel infrastructure does not happen in affluent areas. Pipelines, refineries, and compressor stations don't get placed in the fancy parts of town. They get placed in areas that have historically been subject to economic and political exclusion at the hands of racially discriminatory or economically exploitative policies. Appropriately named "sacrifice zones," these areas came to the public discourse in the 1970s in regard to strip mining that was devastating Appalachian and western communities of poor workers—who, ironically, were often employed by the same industry that was causing them harm.

This flawed form of utilitarianism would continue to dominate economic theory, eventually being used to justify the placement of expanded fossil fuel infrastructure, coal power plants, petrochemical facilities, landfills, and the like. The idea that some communities—usually those that were either too poor or too melanated, too rural or too coastal, too undesirable—needed to be sacrificed for the greater good, or to provide some intrinsic value to society, continues to this day.

In April 2022, a study by joint researchers at the University of California–Berkeley, the University of California–San Francisco, and Columbia University confirmed this. In comparing data on the location of plugged and active oil and gas wells to data from maps demonstrating redlining that occurred under the Home Owners' Loan Corporation, the joint study found that historically redlined communities have nearly 200 percent more oil and gas wells than non-redlined neighborhoods.[11]

How did this happen? Enter the *Underwriting Manual* of the Federal Housing Administration. This document established the process of redlining, saying that "incompatible racial groups should not be permitted to live in the same communities," meaning that loans to Black and immigrant communities could not be insured. This document also went further, with a recommendation that highways would be an effective tool for separating Black and immigrant communities from white neighborhoods.[12]

Beginning in the early 1930s, highways were preferentially constructed through Black and brown communities across the nation. Entire communities where Black and immigrants had found limited opportunities for home ownership were bulldozed for preferred placement of our National Highway System—communities like Freedman's Town, in North Dallas, Texas; West Bellfort in Houston; and West Oakland in California.[13]

These highways served as channels of commerce along which big industry developed, searching for locations to place the polluting assets of their operations in areas that were near these channels of commerce but also did not have the political or financial capital to offer significant resistance. These locations happened to be the same communities—poor communities and communities of color—that, because they were kneecapped in building wealth by being shut out of the homeownership process, were now economically and politically vulnerable. Today, highways are still a significant source of air pollution with communities of color—the same communities that were redlined being disproportionately exposed to vehicular air pollution.

This course of action put in place a feedback loop of inequity, which still exists. It's the reason that today, 68 percent of African Americans live within thirty miles of a coal-fired power plant. It's the reason that African Americans are 75 percent more likely than white people to live in fenceline communities close to oil and gas infrastructure that pollutes our air and changes our climate. It's the reason that, in forty-six US states, people of color live with more air pollution than white people.

This legacy also leads to the development of comorbidities in these communities: higher exposure to air pollution and lower exposures to green space mean higher morbidities from asthma, respiratory disease, and death from extreme heat—*all issues that are made worse with the impacts of the climate crisis.* Thereby this phenomenon—a racially discriminatory policy decision made over a century ago—becomes the norm generation after generation. And the greatest cruelty of all: these same communities that have endured this feedback loop are now hit first and worst by the climate crisis, despite contributing the least to its design. Poor communities, in the United States and around the world, despite paying a higher percentage of their income to meet energy needs, on average, consume less energy per capita than their wealthy counterparts, meaning that they have historically been less responsible for the demand for fossil fuel energy.

This history makes it critical for us to see these communities not as victims, but as communities victimized by intentional bad policy decisions. It is also the reason that, in our efforts to stymie the current ecological crisis, we must infuse, into the core, lenses of climate and environmental justice. Van Jones says it like this:

> The green economy should not just be about reclaiming throw-away stuff. It should be about reclaiming thrown-away communities. It should not just be about recycling things to give them a second life. We should also be gathering up people and giving them a second chance. In other words, we should use the transition to a better energy strategy as an opportunity to create a better economy and a better country all around.[14]

The current and future transition to a clean energy economy represents the emergence of a multi-trillion-dollar market, on the scale of $50 to $131 trillion.[15] Comparatively, the entire Internet of things was valued at $8.9 trillion in 2020. Additionally, a burgeoning mass of political discourse is calling for not only climate investment, but climate investment through a lens of equity that seeks to redress historic harm—and also create opportunity for ownership and self-determination for communities. The latter is especially key: to see the scale of this current transition—which sits on a scale of multiple times of the current entire market for the Internet of things—and to talk about climate justice for frontline communities only through the limited language of their role as a workforce, and not as owners in the emerging economy, is a disparity in and of itself. It is a form of climate apartheid, wherein we delegate these communities into a well-paid underclass that serves as our worker bees and not as our true co-collaborators of the clean energy future.

In my personal activism, I've coined this moment as the moment of 7Cs: the Critical Coalescence of Consciousness and Capital to Combat the

Climate Crisis. In this moment of 7Cs, we must recognize that with this transition comes a once-in-existence chance to begin to redress *some—not all—but some* of the harm that has been historically done to frontline communities, here in the United States and across the globe. This moment is of such a scale that no one silo of society can accomplish a solution on its own: science cannot solve it on its own; government cannot solve it on its own; private finance cannot solve it on its own; grassroots activism cannot solve it on its own; the faith community cannot solve it on its own. It will take all of us, united under a new politic, to meet this moment, with the best that our collective humanity has to offer. In doing so, we secure a future for us all.

EPILOGUE

We face the fierce urgency of now, in which we must change the priorities of a society that claims to be the most powerful in history—technologically, economically, educationally, politically, militarily—or there may not be very much more history. Each of us, at this moment, is met with a moral and generational obligation, not to despair, but to continue to work to effect change; not just to strive for it, but to see it—enacted—if there is to be a future. In doing so, we redeem ourselves and prove ourselves worthy of this great human experiment.

5 GOVERNING SCIENCE, TECHNOLOGY, AND INNOVATION IN HOTTER TIMES

David H. Guston

COLUMNS AND CORALS

This volume begins with a set of assumptions about democracy and climate change. The first set, about politics, holds that democracy has practical and transcendent value that deserves to be sustained. The second set, about science, holds that climate change poses an imminent and existential threat not only to the Earth's natural systems as we know them, but also to these democratic values and their systems. Under these assumptions, the marble columns of our democratic institutions are as threatened by global warming as the corals of our acidifying seas.

Buying into these assumptions as a tentative diagnosis is one thing. Yet what follows is quite a dilemma: global warming is rapidly eroding our democracy, but we must fix our democracy in order to fix global warming. The horns of the dilemma are sharpened with the realization that we are now working *in medias res*. We know the story of how things began in hubris and deception and, even if fate is not guiding our wanderings, then certain path dependencies are. Even if we want to undo effects, political or scientific, we cannot undo causes; we must create new causes for new effects. Our scientists tell us we must act urgently, but learning enough to not repeat past mistakes takes time, and we cherish a constitutional system designed to be skeptical of haste.

Furthermore, recent edits to this story line, not just from climate change but also from COVID-19 and elsewhere, suggest that Americans are as starkly divided on issues related to the authority of science and technology as they are on issues more traditionally considered as "political." For some, it seems, science operates no differently than political parties and advocacy, while for others science is the pole around which politics and policy should revolve. The challenge to liberal democracy is not now limited to how one can live among those of difference conscience, as it was in the seventeenth century, but how one can live among those of a different science!

Science, technology, and innovation—I refer to them jointly as ST&I—are deeply implicated in these threats, to both democracy and climate, as well as to the foundations of their own support and success. It is not just knowledge-based industrial processes that release carbon dioxide and methane to create greenhouse conditions that are to blame, but it is also the organization and ideology of ST&I that hold them aloof from clear accountability to the public good. Nevertheless, ST&I have been part of the diagnostic capability that revealed global warming and understands its profound consequences for the inhabitants of our planet. They are also likely constituents of any cure for or coping with global warming and its sequelae. Beyond their role in climate, ST&I are important factors in quality of life, employment, and economic and cultural opportunity and equality. There are approaches we can develop to minimize their negative impacts and encourage their contributions to diagnosis and coping. Indeed, we will need to find these better ways to govern science, technology, and innovation if we want to save both the columns and the corals.

SUCCESSES AND FAILURES

While this volume proceeds from the above assumptions, this chapter begins with a set of supplementary ones—about ST&I and what we would want from a democratically robust and helpful approach to them, a *governance regime* for ST&I, if you will.

First, we would want our governance regime for ST&I to respect their deep tradition in US life and thought. Our Enlightenment founders thought of themselves as engaged in a science of politics and drew inspiration from Newtonian physics. American invention and innovation in medicine, energy, agriculture, business, and electronics has driven unimaginable material progress. ST&I stocked the arsenal of democracy, initially secured by radar, the proximity fuse, penicillin, and the atom bomb but extending into the digital age with the Internet, GPS, and antimissile systems. And modern science has given us ideals that correspond to and reinforce ideals of liberal democracy, particularly around evidence, deliberation, representation, and the formation of consensus through nonviolent change.

The second assumption is that it would be a mistake to take the first assumption as the whole of the story. The founders often had ridiculous ideas about topics related to ST&I, including race, heredity, and destiny. They are divided from us by fundamental ideas about the atom, the gene, and the mind, as well as by a suite of industrial revolutions that turned water wheels into steam turbines, Franklin's "electric fluid" into wireless power transmission, lithography into computer chips, and mortars into hypersonic missiles. The connection is tenuous at best between technical progress and moral progress, and we substitute the former for the latter at great risk, as much contemporary innovation reproduces earlier inequities. We have misused and abused our technological prowess on the national and global stage, and scientists have too often been too comfortable heeding the hand of power rather than hearing the still, small voice of virtue.

We should aim for a governance regime that learns from the ways that ST&I has contributed to both the successes and failures of the American experiment. It should contribute to sorting through the practical challenges of how to deploy scarce resources in the battle against climate change—including not just what research needs to be done, but also how to extend the results of such research to make our communities safer, healthier, and more resilient in the face of a violently changing planet.

At the same time, the governance of ST&I should strengthen democratic norms and values, and it should explore ways to heal the policy divides over science. We want ST&I that is itself more democratic, that distributes its bounties more equitably throughout the country (and the world), and that puts a vision of public value in the forefront of its aspirations.

INNOVATION AND REPRESENTATION

Finding guidance to turn these assumptions into a concrete agenda is oddly not all that difficult. My preferred source is the political scientist Langdon Winner who, in his 1977 book *Autonomous Technology*, identified technology with legislation. He argued that technology and legislation are aspects of the same kind of structuring of society that we do collectively and through which we all pursue what we think is good in the world. Laws and technologies similarly give shape to the world, operate as parts of systems in themselves and in concert with other systems, and interact importantly with human intentions to facilitate or inhibit our pursuit of our goals.

Once we understand technology in this way, the normative implications are direct and profound: if we believe we need democratic norms and institutions to make laws, then we must similarly believe that we need democratic norms and institutions to make technology, that is, to innovate. In an eighteenth century dominated by Hanovers and Bourbons, the American slogan of "no taxation without representation" was revolutionary. In a twenty-first century dominated by Musks and Thiels, "no innovation without representation" becomes similarly revolutionary.

One core meaning of "no innovation without representation" is that the market is an insufficient forum for the adjudication of decisions around ST&I. We already know this fact through experience with some technologies, wherein the risks of an unregulated system are relatively easy to identify, such as drugs and pesticides. The governance regime for drugs and pesticides and similar products in the United States is known as the risk paradigm. To authorize action, the risk paradigm standardly requires sophisticated scientific and quantitative assessments of the kinds of bad outcomes, or hazards, that can occur and their probability of occurring. While the risk paradigm has notable successes, it has at least three important and relevant problems.

The first problem is that the scope of hazards considered by the risk paradigm is nowhere near as broad as it should be, focusing mostly on easily identifiable medical outcomes like cancer in human beings (although other outcomes have been introduced more recently, including fetal health). We might, for example, have questions about the physical health effects of 5G technologies, emphasized because we would know how to create regulations if we found such effects. We have nevertheless had much discussion but little action on the mental health effects of smartphones, social media, and other digital tools, especially on children, in part because it is much more difficult to measure and establish causes for mental health outcomes (e.g., there is no reasonable animal model for depression and suicide induced by social media).

The case of digital technologies, similar to that of drugs and pesticides, also illustrates a second problem: the real potential hazard may not be limited to the single entity—the single drug, pesticide, or phone—but to the systems of drugs, pesticides, and digital tools that are used in concert and may have interactive effects. The cell phone in an individual user's hand is not so much the challenge as the more complex reality of cell phones and their social media apps being densely networked and reaching a level of saturation that changes people's behaviors (e.g., while driving) and expectations of others' behaviors (e.g., availability for work). Potentially malign effects like the revelation of data or metadata that even if anonymous can be turned back on users, or the viral spread of misinformation to the detriment of fact-based decisions at individual and social levels, can spread as a function of the network.

A third problem is that ST&I moves very quickly, and increasingly so. Some argue that the law lags behind changing technologies and thus we need a more permissive legal response. But that comparative speed is not an immutable quality of ST&I. We not only choose to decelerate the creation of new laws through institutional checks and balances and requirements for administrative procedures, for example, but we also choose to accelerate innovation by stoking it with direct funding for academic research, providing direct funding and tax credits for corporate research, and valorizing it above the less "sexy" but similarly important policy investments like standards and material investments such as infrastructure maintenance. I would rather emphasize that the risk paradigm drives law to make decisions using scientific data, which take time to generate. New and emerging technologies, including precautionary action to mitigate or adapt to ongoing climate change, usually lack data not just about the hazards they could cause but more so about the probability of those hazards manifesting.

These reasons are not necessarily sufficient to jettison the risk paradigm in its entirety. But we clearly need a governance regime that addresses a broader set of potential hazards, including the social risks, of ST&I. We also need better ways of grappling with the systemic aspects of ST&I's risks, as well as better ways to match the timing of data, decisions, and deployment of ST&I. Adopting a precautionary principle could be somewhat helpful here, but I deeply appreciate philosopher Jean-Pierre Dupuy's take on precaution, which, he argues, still depends on a quantitative assessment of how much is enough to act. In this way of thinking, precaution is therefore still anchored to the risk paradigm.

DIVERSITY AND DIALOGUE

So, one program to advance "no innovation without representation," focusing mostly on the innovation side, might be to facilitate collective decision-making about ST&I that does not rely on the market and substantially improves on the risk paradigm in scope, breadth at the systems level, and timing of knowledge production and intervention (I return to some details about such a program below). A second program to advance "no innovation without representation" is to focus on the representation side, literally to create more representative ways of innovating.

Several approaches could contribute to such a program. The first is making the human diversity of the ST&I ecosystem representative of the human diversity in the country. Diversifying the students and workforce in STEM (science, technology, engineering, and mathematics) has been a long-term goal that has seen modest successes and spectacular failures. Expanding representation is not just about women and racial minorities earning STEM degrees, however—it is also about researchers and teachers, entrepreneurs and financiers, philanthropists and program officers, regulators and public affairs professionals, inspectors and advocates, legislators and judges. We need to diversify the entirety of the ST&I ecosystem, which touches all our contemporary institutions, so that the people driving the ideas of innovation bring more diverse visions of what ST&I can do, and who it can be done for, than the path that we have been on.

As much as new and different people need to inhabit the ST&I ecosystem, the structural elements of that system still may not well serve the aspirations of the diverse people in it. As Churchill said of the structuring effects of architecture, "We shape our buildings; thereafter they shape us." Since the ST&I ecosystem has been shaped by a narrow group of people with narrow goals, it impresses those systemic perspectives—through engineering curricula, accreditation requirements, and employment opportunities, for example—on all its participants, however diverse they may be. Moreover, ST&I is seldom among the most salient issues for most of the many institutions touched by the ST&I ecosystem—even if ST&I permeates many other salient issues, like the economy, education and, of course, climate. "No innovation without representation" therefore fails miserably as an operating concept for even those institutions designed to be representative.

Making ST&I more salient means strengthening the ways we inform or educate both the public and the decision-makers who are, ideally, responsive to them. The mechanism of that information or education cannot simply be, however, a unidirectional and monotonal communication by scientists and technologists that ST&I is good and therefore good decisions should follow, as in the first assumption about ST&I above. It should rather be built around a more complete dialogue that motivates the breadth and ambiguity of the above assumptions. In essence, the dialogue that animates the ST&I ecosystem should be responsive to the wonderful observation by Jeff Goldblum's character, Dr. Ian Malcolm, in the 1993 movie *Jurassic Park*: "Scientists were so preoccupied with whether or not they could that they didn't stop to think if they should."

A more academic version of Dr. Malcolm's question exists in the related fields of responsible research and innovation (RRI) and public interest technology (PIT). Both RRI and PIT attempt to engage with the public and private sector portions of the ST&I ecosystem to promote this more complete dialogue. The European Commission (EC) and a few northern and western European nations have elaborated RRI most substantially, implementing it in research programs of the EC and in the Engineering and Physical Sciences Research Council of the United Kingdom and elsewhere. PIT has emerged in the United States in a less organic and more self-conscious way, largely through the efforts of New America and funding from the Ford Foundation and oriented toward a growing network of now roughly four dozen colleges and universities. The PIT movement is modeled after the development of public interest law, which Ford envisioned and bankrolled two generations ago. Indeed, the need for PIT after public interest law follows Winner's legislation-technology identity at the heart of "no innovation without representation." Many of the programs developed by member institutions of the PIT University Network attempt to expand access to the ST&I ecosystem for women and minorities, engage with policy-makers and community members about the deployment of new and emerging technologies in the public interest, and support reflexive conversations across multiple disciplines on the meaning and purpose of innovation.

ASSESSMENT AND PARTICIPATION

If the ST&I ecosystem were more representative demographically, if the people who inhabited its institutions and the people they represented found ST&I more salient, and if they all were further supported by a community of like-minded thinkers supporting technology in the public interest, our traditional institutions might govern ST&I better. It might be more protective of people's lives and interests, and it might be both more innovative and more public-spirited in its diversity. But unless and until we accomplish that immense task, we might also consider supporting additional institutions and pathways that sometimes sidestep existing institutions but still reinforce a supportive, reflective, and deeply American approach to governing ST&I.

One starting place is with individual members of the broader political community itself. Governing ST&I has, from the perspective of citizens or laypeople, been largely limited to choosing to buy or not to buy what the market might offer, and perhaps choosing among political representatives who have little knowledge of and pay little attention to ST&I issues. There are opportunities for laypeople to engage in more in-depth self-governance vis-à-vis ST&I and to create, thereby, more influence and develop more trust.

One opportunity is direct participation in ST&I decision-making. Beyond regulation, one of the ways in which government has sought to supplement the market, and inform political representatives about ST&I, is through technology assessment—scientifically and technically sophisticated, future-oriented policy analysis. The United States once institutionalized such work in an Office of Technology Assessment (OTA) that served the Congress from 1972 to 1995. OTA wrote (usually) book-length technology assessments through the work of in-house experts, hired contractors, and expert advisory groups.

For much of its existence, OTA kept to a tenuous bipartisan and bicameral balance. In 1994, however, Representative Newt Gingrich (R-GA) promised, in his Contract with America, to put OTA on the chopping block should his party take control of Congress in that year's midterm elections. The Republicans did, and in 1995 Congress defunded OTA. Despite his hostility to OTA, partially based in an idea that politically biased professionals and intellectual elites inappropriately intercede between experts and decision-makers, Gingrich regarded himself a futurist of sorts. He hoped that scientists and members of Congress might have closer communications without OTA's kind of intermediaries. Gingrich's right-wing populism did not produce the only critique of OTA as elitist; a left-populist critique held that the voices of ordinary people could help inform political debate even around scientific and technical issues.

Since the demise of OTA, a more participatory technology assessment (pTA) has in fact arisen, both in the United States and globally. In 1997, the small not-for-profit Loka Institute conducted the first "consensus conference" or "citizens' panel" in the United States, modeled after a process pioneered by the Danish Board of Technology. Focused on the issue of telecommunications and the future of democracy, a dozen or so laypeople from the Boston area deliberated with a small group of experts and wrote their own public report. It became a model for other experiments in locally conducted pTA, and as digital tools for dialogue and deliberation emerged, others experimented with computer-mediated panels. In 2008, the Center for Nanotechnology in Society at Arizona State University used such tools to conduct the first National Citizens' Technology Forum (NCTF) in the United States, connecting small-scale, face-to-face deliberations in six sites across the country through synchronous and asynchronous keyboard-to-keyboard deliberations.

The next year, the Danish Board organized the first World Wide Views process, on global warming, linking numerous sites across thirty-eight countries on all six inhabited continents—nearly four thousand laypeople in total. The expertise developed by the NCTF was crucial to US participation in the WWViews on Global Warming, which reported its results at the 2009 UN Climate Change Conference in Copenhagen (COP15). The pTA team at ASU then joined with the Loka Institute, the Boston Museum of Science, the Woodrow Wilson International Center, and a citizen-science group called SciStarter to create the Expert and Citizen Assessment of Science and Technology (ECAST) Network; in 2012, ECAST served as the formal coordinator for four US cities to participate in the World Wide Views on Biodiversity, which reported to the Eleventh Council of Parties of the UN Convention on Biodiversity.

ECAST and the ASU pTA team, led by Mahmud Farooque, have continued to pioneer and even normalize their approach to pTA. Their projects over the last decade involve continued collaborations with WWViews (on climate and energy, and on oceans and seas) and sponsorship from US executive agencies such as NASA, NIH, the Department of Energy, and the National Oceanic and Atmospheric Administration, as well as private sponsors on topics like driverless cars (also with a global audience in conjunction with the Paris-based Missions Publiques), geoengineering, and human gene editing after CRISPR.

CITIZENS AND SCIENCE

The outcomes that yoke these various pTA events is their ability to bring together diverse people with no previous, specific knowledge or experience in an ST&I topic, have them learn about it collectively, discuss important related public issues, and have their voices travel as input to those more traditional institutions. But even as far as pTA goes in democratizing decision-making about ST&I, one can look further still at much more democratized processes of doing ST&I, including citizen science and DIY science and collectives.

Citizen science refers to the contributions to knowledge creation that laypeople, not formally trained in scientific investigations, can make. Before the term scientist was coined in the 1830s, and many scientists became professionalized through the nineteenth century, natural philosophers and natural historians were often amateurs—making observations, gathering specimens, and so forth as an avocation rather than a professional activity. After science became a vocation with the rise of professions and research universities, along with consequent specialization and methodological and instrumental sophistication, the practice of science became vastly more distant from ordinary life.

The amateur has never quite departed, however, especially in areas akin to natural history. For instance, among the earliest and most robust examples in contemporary research of citizen scientists providing important data in conjunction with professionals is from ornithology, which over several decades mobilized backyard birders to observe and catalogue the identity and numbers of nesting, wintering, and migratory birds. While birders have simple and accessible instruments like feeders and binoculars, the addition of digital tools like apps that can identify birds by their songs means a further democratization of citizen-science expertise for data gathering. In other fields, including astronomy and environmental science, such digital tools as phone cameras and distributed computer programs and computing power have allowed citizen scientists to contribute to more complex discoveries.

Because of the divide between professionals and amateurs, citizen science needs dedicated mechanisms to bring the two groups together. The Cornell Lab of Ornithology has served in this role between professional ornithologists and amateur birders, and it has also investigated and documented through social science and evaluation research the quality and contributions of citizen science. Another digital tool, the website SciStarter (named after KickStarter), matches aspiring citizen scientists with formal research projects that can use their assistance.

Citizen science received a huge boost from Congress in the form of the Crowdsourcing and Citizen Science Act, which was incorporated into the bipartisan American Innovation and Competitiveness Act of 2017. The act expressed the sense of Congress that crowdsourcing and citizen-science projects have a number of additional unique benefits, including accelerating scientific research, increasing cost effectiveness to maximize the return on taxpayer dollars, addressing societal needs, providing hands-on learning in STEM, and connecting members of the public directly to federal science agency missions and to each other. Granting federal science agencies the direct, explicit authority to use crowdsourcing and citizen science will encourage its appropriate use to advance federal science agency missions and stimulate and facilitate broader public participation in the innovation process, yielding numerous benefits to the federal government and citizens who participate in such projects.

Despite these potential benefits, a common mode of conducting citizen science is to have laypeople act, in essence, as volunteer research labor on a project designed by elite scientists for their own goals. SciStarter has begun collaborating with public libraries to help promote citizen science in areas where more sophisticated tools or training might be necessary to conduct the inquiries, giving laypeople greater opportunity to perform ST&I on their own agendas. Science shops, as they were known in an earlier generation, are places where universities make their ST&I capacity open to direction from community members, asking questions that are important to them, for example, about local environmental conditions like air or water quality. In their 2019 book about citizen science, *Science by the People*, sociologists Aya Kimura and Abby Kinchy observe several grassroots organizations operating in a citizen-science or science shop mode. Although the organizations suffer from competing logics, they can also be effective and politically relevant ways of communicating authentic, scientifically informed voices.

Connections are also emerging among citizen science and maker spaces, do-it-yourself (DIY) biology, and other movements that demonstrate again the sometimes seamless connections within ST&I activities. GenSpace, in Brooklyn, New York, is a collective of DIY biologists who model the connection between politics and ST&I in yet another way, by foregrounding "core values" and a code of conduct that goes beyond traditional scientific norms by integrating them more directly with democratic social norms. The strategic aim for such DIY collectives is, according GenSpace's strategic plan, a world in which "people feel a sense of identity, belonging, and familiarity with emerging life sciences, especially for those who don't usually see themselves reflected in traditional scientific spaces. Science is demystified."[1]

ANTICIPATION AND GOVERNANCE

For all the ways that citizen science, science shops and collectives like GenSpace might effectively merge ST&I with democratic virtues, and to the extent that they flange up with decision-making as well, they are still largely reliant on a system committed to the risk paradigm—with all its flaws—for influence. An important and distinctive supplement to the risk paradigm exists. This alternative is more open to qualitative and systemic data, or even missing data; it is more welcoming of diverse voices and more attuned to broader perspectives on public values and public interest; and most importantly perhaps, it is well-attuned to aspirations for moral as well as technical progress. This perspective on ST&I governance is called *anticipatory governance*, and it has evolved over the last decade or so as a way of meshing the societal need for innovation, on one hand, with different ideas of what is desirable from innovation, on the other. It provides a collective opportunity to answer Dr. Malcolm's question about whether we should innovate and how.

Having been developed only recently, the idea of anticipatory governance boasts some deep intellectual roots. One of these roots taps visions of interdisciplinary collaboration, and particularly the important role of the social sciences as partners or collaborators with the natural sciences and engineering. In the late 1940s, Detlev Bronk, one of the founders of the interdisciplinary field of biophysics and a president of both Johns Hopkins University and the National Academy of Sciences, was an early visionary in this regard. Despite being an elite scientist, Bronk broke with his colleagues when testifying before Congress in favor of the proposal to create a National Science Foundation (NSF) after World War II. Contrary to prevailing opinion, Bronk believed that a new NSF should fund social science so that "competent social scientists should work hand-in-hand with natural scientists, so that problems may be solved as they arise, and so that many of them may not arise in the first instance."[2] Bronk touches not just the interdisciplinary heart of anticipatory governance but also its connection to problem-solving in an anticipatory manner.

Interdisciplinarity is now all the rage at universities, although it is still more aspired to than implemented. One important effort, developed by Erik Fisher at ASU, is "socio-technical integration research." STIR, as it is known, embeds humanists and social scientists as participant-observers and interlocutors in natural science and engineering laboratories. What they accomplish is very much in the spirit of Bronk's envisioned collaboration, if not precisely its "hand-in-hand" equity. STIR has spread into many countries and into private-sector as well as academic research laboratories, where it often identifies or instigates sometimes subtle, sometimes profound shifts in perspective, methods, or strategy. The protocol operates only at the scale of the laboratory, and only in several dozen cases at that, but it nevertheless shows great potential to increase the ability of scientists and engineers to reflect on their goals and their plausible successes and failures, and to consider their own laboratories as sites of the governance of emerging ST&I.

A second root of anticipatory governance tapped the work of sociologist and popular futurist Alvin Toffler, who developed "anticipatory democracy" as an antidote to the *Future Shock* he diagnosed in his 1970 book. For Toffler, the democratic part of anticipatory democracy meant the tradition of popular democracy in New England town meetings—not very far from the consensus conferences or citizens' panels described above. The "anticipatory" part meant using predictive techniques, particularly demographic and economic forecasting with then new-fangled computers, to inform popular deliberations about likely future states of the world.

This computer-based modeling tradition continues in state-level politics where real or gamified versions of participatory budgeting and redistricting, much as Toffler envisioned, have thrived. It also continues, of course, in the context of climate change, with global circulation models and the forecasts from the IPCC and other groups about the probabilities of future climatological conditions. But such modeling—akin to the risk paradigm—works best when model-builders have cascades of data and refined causal knowledge to make predictions precise enough to be useful.

Anticipation is not meant to take part in the hubris of predicting or forecasting or modeling a single future. Rather, as envisioned and developed by ASU's Cynthia Selin, it is meant to explore plausible futures for guidance into how to begin, now, to construct pathways to normatively attractive ones and determine whether we might still be on those pathways as we proceed. A central technique of this mode of anticipatory governance is scenario development, which is grounded in more qualitative, dialogic, and even narrative approaches to describe events, situations, and systemic relationships that are plausible in some future time. Like the STIR protocol above, scenario development workshops designed for scientists and engineers—with the participation of additional interdisciplinary scholars and prospective downstream users of innovations that the research might yield—have proven effective in creating both tactical and strategic shifts in research agendas.

While the scenario development approach was pioneered in the private sector in the mid-1960s, prior to widespread computerization, it has resurfaced over the last decade or so into broader acceptance and application—including endorsement by NSF as an important approach to governing emerging technologies. Indeed, a new field of "anticipation" or "anticipation studies" may be developing through recent efforts of the Social Science Research Council in the United States, and an international group of scholars and practitioners that held its fourth International Conference on Anticipation at ASU in November 2022.

OBJECTIONS AND REPLIES

Chapters in this book, and critics elsewhere, have noted that some elements of the agenda articulated above can work against the overarching goal of innovating sufficiently quickly to meet climate deadlines. (Indeed, it might do our motivation good to recapture the original, morbid nineteenth-century meaning of the word "deadline"—a mark in the landscape beyond which a prisoner would be shot and killed.) On the topic of small-scale democracy, for example, many neighborhoods embrace not-in-my-backyard (NIMBY) as enthusiastically for renewable energy technologies like wind turbines as they do for fossil fuel power plants, refineries, and other polluters. But what this version of NIMBY also shows is that people yearn for a level of engagement and influence over their environment at scales relevant to them. That yearning can be a resource for democratic leadership in these hotter times, as the consequences of climate change can be rendered visible in communities and not just through general circulation models and mean global temperatures (in Celsius, no less!).

Another objection to the list of democratic alternatives for ST&I that I have offered here is that they are too small in scale, and that grander, more sweeping constitutional-level changes are needed. Harvard's Sheila Jasanoff argues this claim in her 2016 book, *The Ethics of Invention*. Just as I would not reject many of the grander constitutional interventions Jasanoff imagines, however, she would not abandon these reforms around anticipatory governance. And given the existing political divisions that form part of the premise for this volume's inquiry, the interventions into public engagement and participation, interdisciplinarity, and anticipation are already bubbling up through society, and thus they seem better placed to make the kind of shorter-term impact that more consensus-based activities can have. The grander efforts may well be impossible without this groundwork.

Potentially more challenging is the question of whether there are boundaries of technological or other complexity to the kind of democratic control that "no innovation without representation" implies. This battle harks back to Socrates, his argument for guardianship, and the "noble lie" of the inherent superiority of the class of guardians. In the twentieth century, the debate between democracy and guardianship was perhaps most intensely joined on the governance of nuclear weapons and the creation of new institutions like the congressional Joint Committee on Atomic Energy and the Department of Energy, and the evolving doctrine about war powers and the ability of the president of the United States to unilaterally engage in hostilities, including authorizing a nuclear strike. For all intents and purposes, POTUS seems a "thermonuclear monarch," as Harvard's Elaine Scarry has written.

If climate change is an existential threat akin to thermonuclear war, have I still not even contributed to solving the problem posed by this volume?

DEMOCRACY AND REFORM

The extent to which decision-making about nuclear weapons, and thus nuclear war, is a good analogy for decision-making about climate change might make an excellent essay in itself. In brief, such an essay would attempt to determine whether there exists a category of technology, pertinently here nuclear weapons and technologies that address climate, that, by virtue of inherent characteristics of the technology and its context of use, precludes democratic governance. Langdon Winner, whose work inspired the "no innovation without representation" framing I have used here, thought that nuclear power plants fit this category, by dint of their technical complexity, scale of operation, and national security requirements. In his view, only a large government could operate a nuclear power plant safely, and then only through largely authoritarian means. For nuclear weapons and climate technologies, the key characteristics, I think, are arguments about technological complexity; the need for prompt action; and the dire, global nature of consequences, rightly or wrongly decided.

Such an essay, roughly, exists. Published in 1985 by Yale political theorist Robert Dahl, *Controlling Nuclear Weapons* pits versions of guardianship against democracy. Dahl asks the poignant question, "Have we then reached an inherent limit on democracy, a fundamental and inescapable defect? If so, are we rationally obliged, no matter how much our deepest feelings may tug us away from the dismaying conclusion, to commit ourselves to a nondemocratic alternative, at least over some set of public policies?"[3] In a thorough treatment that neither caricatures guardianship nor idealizes democracy, Dahl articulates how a blend of moral competence and instrumental competence is required for public decision-making on the part of prospective guardians and citizens alike. But guardians, he argues, have little claim on instrumental competence particularly as the specialization of knowledge means that, at best, potential guardians have deep knowledge in one narrow domain, generalist knowledge across other domains, and no necessary advantage in moral competence at all. He concludes that a political system committed to the equal intrinsic worth and personal autonomy of all its participants would perform better at understanding and acting on the public good than would a guardianship.

To arrive at a democratic political system that could handle the complexity and consequence of nuclear weapons—Dahl mentions other technologies as well as environmental degradation in passing, but does not mention climate change, perhaps as the first IPCC did not report until 1990—Dahl offers three, interrelated reforms: The first reform, around public learning, seeks information that is easily and universally accessible, in an appropriate level and form, and accurately reflects the best available knowledge. The second reform seeks direct citizen participation to influence the agenda of existing elites and quasi-guardians. To represent publics that cannot participate directly, the third reform seeks to incorporate public opinion through the constitution of what he calls a "minipopulus" (or what here is called a consensus conference). Indeed, if you read just a little between the lines, for instance, in Dahl's own example of using computer visualization to assist in representing what a planned building will actually look like at scale and integrated into its future environment, you begin to see how closely his suggestions and anticipatory governance map onto each other.

Another version of an essay about ST&I and democracy might talk about particular policy areas such as intellectual property or data policy or public spending on scientific research and development, particularly at universities. But a close read of this essay, I think, will yield my suggestions, implicit or explicit, about such topics. What follows for data policy, for example, if we value citizen science and easily and universally accessible information? (And as I complete this essay, the Biden administration's Office of Science and Technology Policy has just announced a new and more expansive open-access policy for federally sponsored research.) Even for cross-national and global considerations of ST&I—for example, geoengineering techniques in which one country's fairly inexpensive actions could influence not just their immediate neighbors but the entire globe—the opportunities offered by the World Wide Views process points a way to creating a global "minipopulus."

Dahl casts his reforms as "quasi-utopian," and yet they exist in many corners of today's complex society. One of the strengths of democratic governance is indeed that society. When even our democratically structured institutions are stuck, when they fail to encode and enact the will of the people, the people still have other places to go for satisfaction. If they can find avenues for participation, and opportunities to collaborate and to imagine futures together, they can reconstitute themselves in preparation for reconstituting their institutions, rather than yielding them to the forces that would tear them down, column and coral together.

6 CONFRONTING CLIMATE CHANGE IN EXTREMELY ONLINE TIMES

Holly Jean Buck

Burning hills and glowing red skies, stone-dry riverbeds, expanses of brown water engulfing tiny human rooftops. This is the setting for the twenty-first century. What is the plot? For many of us working on climate and energy, the story of this century is about making the energy transition happen. This is when we completely transform both energy and land use in order to avoid the most devastating impacts of climate change—or fail to.

Confronting authoritarianism is even more urgent. About four billion people, or 54 percent of the world, in ninety-five countries, live under tyranny in fully authoritarian or competitive authoritarian regimes.[1] The twenty-first century is also about the struggle against new and rising forms of authoritarianism. In this narration, the twenty-first century began with a wave of crushed democratic uprisings and continued with the election of authoritarian leaders around the world who began to dismantle democratic institutions. Any illusion of the success of globalization, or of the twenty-first century representing a break from the brutal twentieth century, was stripped away with Russia's most recent invasion of Ukraine. The plot is less clear, given the failure of democracy-building efforts in the twentieth century. There is a faintly discernable storyline of general resistance and rebuilding imperfect democracies.

There's also a third story about this century: the penetration of the Internet into every sphere of daily, social, and political life. Despite turn-of-the-century talk about the Information Age, we are only beginning to conceptualize what this means. Right now, the current plot is about the centralization of discourse on a few corporate platforms. The rise of the platforms brings potential to network democratic uprisings, as well as buoy authoritarian leaders through post-truth memes and algorithms optimized to dish out anger and hatred. This is a more challenging story to narrate, because the setting is everywhere. The story unfolds in our bedrooms while we should be sleeping or waking up, filling the most quotidian moments of waiting in line in the grocery store or while in transit. The characters are us, even more intimately than with climate change. It makes it hard to see the shape and meaning of this story. And while we are increasingly aware of the influence that shifting our media and social lives onto big tech platforms has on our democracy, less attention is devoted to the influence this has on our ability to respond to climate change.

Think about these three forces meeting—climate change, authoritarianism, the Internet. What comes to mind? If you recombine the familiar characters from these stories, perhaps it looks like climate activists using the capabilities of the Internet to further both networked protest and energy democracy. In particular, advocacy for a version of "energy democracy" that looks like wind, water, and solar; decentralized systems; and local community control of energy.

In this essay, I would like to suggest that this is not actually where the three forces of *rising authoritarianism* x *climate change* x *tech platforms domination* leads. Rather, the political economy of online media has boxed us into a social landscape wherein both the political consensus and the infrastructure we need for the energy transition is impossible to build. The current configuration of the Internet is a key obstacle to climate action.

The possibilities of climate action exist within a media ecosystem that has monetized our attention and that profits from our hate and division. Algorithms that reap advertising profits from maximizing time-on-site have figured out that what keeps us clicking is anger. Even worse, the system is addictive, with notifications delivering hits of dopamine in a part of what historian and addiction expert David Courtwright calls "limbic capitalism."[2] Society has more or less sleepwalked into this outrage-industrial complex without having a real analytic framework for understanding it. The tech platforms and some research groups or think tanks offer up "misinformation" or "disinformation" as the framework, which present the problem as if the problem is bad content poisoning the well, rather than the structure itself being rotten. As Evgeny Morozov has quipped, "Post-truth is to digital capitalism what pollution is to fossil capitalism—a by-product of operations."

A number of works outline the contours and dynamics of the current media ecology and what it does—Siva Vaidhyanathan's *Antisocial Media*, Safiya U. Noble's *Algorithms of Oppression*, Geert Lovink's *Sad by Design*, Shoshana Zuboff's *Surveillance Capitalism*, Richard Seymour's *The Twittering Machine*, Tim Hwang's *Subprime Attention Crisis*, Tressie McMillan Cottom's writing on how to understand the social relations of Internet technologies through racial capitalism,[3] and many more. At the same time, there's reasonable counter-discussion about how many of our problems can really be laid at the feet of social media. The research on the impacts of social media on political dysfunction, mental health, and society writ large does not paint a neat portrait. Scholars have argued that putting too much emphasis on the platforms can be too simplistic and reeks of technological determinism; they have also pointed out that cultures like the United States' and the legacy media have a long history with post-truth.[4] That said, there are certainly dynamics going on that we did not anticipate, and we don't seem quite sure what to do with them, even with multiple areas of scholarship in communication, disinformation, and social media and democracy working on these inquiries for years.[5]

What seems clear is that the Internet is not the connectedness we imagined. The ecology and spirituality of the 1960s, which shaped and structured much of what we see as energy democracy and the good future today, told us we were all connected. Globally networked—it sounds familiar, like a fevered dream from the 1980s or 1990s, a dream that in turn had its roots in the 1960s and before. Media theorist Geert Lovink reflects on a 1996 interview with John Perry Barlow, Electronic Frontier Foundation cofounder and Grateful Dead lyricist, in which Barlow was describing how cyberspace was connecting each and every synapse of all citizens on the planet. As Lovink writes, "Apart from the so-called last billion we're there now. This is what we can all agree on. The corona crisis is the first Event in World History where the internet doesn't merely play 'a role'—the Event coincides with the Net.

There's a deep irony to this. The virus and the network . . . sigh, that's an old trope, right?"[6] Indeed, read through one cultural history, it seems obvious that we would reach this point of being globally networked, and that the Internet would not just "play a role" in global events like COVID-19 or climate change, but shape them.

What if the Internet actually has connected us, more deeply than we normally give it credit for? What if the we're-all-connected-ness imagined in the latter half of the twentieth century is in fact showing up, but manifesting late, and not at all like we thought? We really are connected—but our global body is neither a psychedelic collective consciousness nor a infrastructure for data transmission comprising information packets and code. It seems that we've made a collective brain that doesn't act much like a computer at all. It runs on data, code, binary digits—but it acts emotionally, irrationally, in a fight-or-flight way, and without consciousness. It's an entity that operates as an emotional toddler, rather than with the neat computational sensing capacity that stock graphics of "the Internet" convey. Thinking of it as data or information is the same as thinking that a network of cells is a person.

The thing we're jacked into and collectively creating seems more like a global endocrine system than anything we might have visualized in the years while "cyber" was a prefix. This may seem a banal observation, given that Marshall McLuhan was talking about the global nervous system more than fifty years ago. We had enthusiasm about cybernetics and global connectivity over the decades and, more recently, a revitalization of theory about networks and kinship and rhizomes and all the rest. (The irony is that with fifty years of talk on "systems thinking," we still have responses to things like COVID-19 or climate that are almost antithetical to considering interconnected systems—dominated by one set of expertise and failing to incorporate the social sciences and humanities). So—globally connected, yet divided into silos, camps, echo-chambers, and so on. Social media platforms are acting as agents, structuring our interactions and our spaces for dialogue and solution-building. Authoritarians know this, and this is why they have troll farms that can manipulate the range of solutions and the sentiments about them.

The Internet as we experience it represents a central obstacle to climate action, through several mechanisms. Promotion of false information about climate change is only one of them. There's general political polarization, which inhibits the coalitions we need to build to realize clean energy, as well as creates paralyzing infighting within the climate movement about strategies, which the platforms benefit from. There's networked opposition to the infrastructure we need for the energy transition. There's the constant distraction from the climate crisis, in the form of the churning scandals of the day, in an attention economy where all topics compete for mental energy. And there's the drain of time and attention spent on these platforms rather than in real-world actions.

Any of these areas are worth spending time on, but this essay focuses on how the contemporary media ecology interferes with climate strategy and infrastructure in particular. To understand the dynamic, we need to take a closer look at the concept of energy democracy, as generally understood by the climate movement, and its tenets: renewable, small-scale systems, and community control. The bitter irony of the current moment is that it's not just rising authoritarianism that is blocking us from good futures. It's also our narrow and warped conceptions of democracy that are trapping us.

HOW CHERISHED NOTIONS OF "ENERGY DEMOCRACY" ARE LEADING THE CLIMATE MOVEMENT ASTRAY

The form of democracy we have now—carbon democracy, as Timothy Mitchell observes—was born along with fossil fuels. "Coal made possible the emergence of mass democracy, and oil set its limits," Mitchell explains.[7] Coal workers could make political demands because of their influence on energy production. Oil, as a liquid commodity, is spread out, and the power to exert pressure lies with the companies and state-owned companies that control it. These forms of carbon democracy are also conditioning our thinking about energy democracy in different ways. We've developed concepts of "energy democracy" in response to those forms of power that oil companies wield—energy democracy becomes small-scale and locally controlled, and renewable. The result is that the activist conception of energy democracy often has nonelected community groups standing in for a democratically elected state, and also fails to grapple with the large-scale energy projects we will need.

Let's talk first about community control. Energy democracy, as a social movement, usually involves an imperative for working at the scale of communities. Up to this point, people have been generally left out of decision-making about new technologies and infrastructures. Hence there is a renewed push to center "the community" or "what communities want." "The community" is the object of focus, rather than talking about areas that vote, such as municipalities or counties. It's often expressed as a form of associative democracy, which centers on voluntary associations, as an alternative to liberal individualism as well as socialist collectivism.[8] One problem with this is that "community" is ill-defined. Anyone can claim to represent the community. Social scientists and development practitioners have been wary of this for decades, observing how monolithic conceptions of "community" mask power relations, for example around gender. But "one Facebook group doesn't want this new geothermal plant" can quickly become "the community doesn't want this geothermal plant." The former is dismissible; the latter has a sacred feel to it. Yet both of these sensibilities are extrademocratic in that they leave out all the people who haven't even heard of the issue, allowing for the possibility of motivated special interest groups to determine the future. Demanding community control is different than demanding *public* control or *public* ownership, and risks becoming a stand-in for these demands.

However, it is inconceivable to build the large-scale infrastructure involved without the direction, coordination, and financing capacities of the state. There are practical problems that are not going to be dealt with through grassroots bottom-up decentralized actions directed by local communities. These include things like figuring out how to maintain appropriate pressure in pipelines as demand for oil and gas declines, figuring out what sort of scale of CO_2 pipelines are needed and planning for them, financing and operating new nuclear power facilities, and funding just transition plans for whole fossil-dependent communities. While some of this can be done by state or even local governments, much of it will need to be decided on and coordinated in much larger structures.

Why are large-scale technologies so necessary? It's moderately possible to describe what it would take to zero out emissions in technical terms, though very few research or policy roadmaps extend to 2100. Partly, this is because we don't know what technologies will be available then. The technical challenges will be immense. But generally, the picture looks something like this: (1) Electrify the power sector using renewables, building out wind and solar, nuclear, and transmission capacity. (2) Produce sustainable fuels, including biofuels, green hydrogen, and synthetic fuels. (3) Decarbonize industry using electricity, hydrogen, and carbon capture and storage. (4) Electrify the built environment and transportation with heat pumps, electric vehicles, and electrified mass transit. (5) Conserve forests and practice conservation agriculture. (6) Improve waste management and build a true circular economy. These actions are described in reports like the IPCC's Working Group III report, *Mitigation of Climate Change*, Princeton's Net-Zero America study, and popular books like *Speed and Scale* by John Doerr and *How to Avoid a Climate Disaster* by Bill Gates. For shorthand, let's call these the conventional wisdom about decarbonization.

Following this conventional wisdom, the path toward decarbonization goes through some things that many environmentalists don't like. Energy systems that are 100 percent renewable face real challenges in terms of energy storage, grid reliability, and the land and materials required. Industrial decarbonization without large-scale, centralized sources of heat like green hydrogen, nuclear, or biomass and capture of process emissions via carbon capture and storage is technically challenging and has not been comprehensively mapped. Moreover, it's not clear what trade-offs the public is willing to make for a 100 percent renewable system compared to an energy system with elements of nuclear, hydrogen, biomass, and carbon capture and storage, which may be cheaper and less land-intensive.

There's an unspoken question beneath the surface: Do climate advocates really want democratic control of the energy system? Are climate advocates prepared for people choosing pathways or technologies they don't like? Will the allegiance be to democracy, or to particular technological regimes? Will we further erode democracy in fear of cheap energy populism, or with the argument that these democratic structures aren't working to represent the vulnerable? Can we risk weakening democracy for these reasons when basic democratic institutions are already under threat from other forces?

A common rejoinder to these tough questions is simply that the conventional wisdom described above is wrong. The models are biased, and if we just had degrowth, or behavioral change, then we wouldn't need nuclear or carbon capture or hydrogen. People who wish to read about these debates can consult Jason Hickel's book *Less Is More*, Giorgios Kallis and colleague's book *The Case for Degrowth*, the new book *The Future Is Degrowth: A Guide to a World beyond Capitalism*, and Leigh Phillips's *Austerity Ecology and the Collapse-Porn Addicts*, and *The Ecomodernist Manifesto*. "Degrowth" is too poorly defined in most popular discourse to be a coherent proposal for how to navigate the century, though quantitative projections and policy proposals at the level of detail of the conventional wisdom would be very welcome. "Behavioral change" and "demand reduction" are clearly going to be necessary in some areas, though again, how this behavioral change is induced in a democracy is not clear, as the incentives-and-penalties approaches to shaping behavior appear unlikely to be implemented by voters at the scale needed to make a climate-significant difference. Degrowth is also unhelpful as a blanket position; there are some things that will need to degrow, and many others that need to grow. But from the data we have, it is clear that clean energy abundance is going to require it all: nuclear, carbon capture, hydrogen, geothermal, energy efficiency, reforestation, behavioral change like opting for less beef, alternatives to fertilizer produced through the Haber-Bosch process, and probably some things we haven't invented yet.

The debates on degrowth versus large-scale technological transformation are dead ends. They are shaped by and manufactured for the platforms, one of many false binaries promoted for engagement and corporate problems. The platforms lead us toward claims about what's possible that wander into post-truth territory. These platforms drive not only two-party polarization, but also micropolarizations within groups. The structure makes it so that every event is divided into opposing takes, leading to animosity between people who are essentially on the same side of addressing climate change. This also bolsters the style of politics that is oriented toward blocking infrastructure, which happens on both sides of the political spectrum. Granted, these tendencies were there before the Internet, but under the outrage-industrial complex, the emotion and action is supercharged. In the United

States, local governments have enacted policies to block or restrict renewable energy facilities in nearly every state, with at least 121 local policies and 204 contested renewable energy facilities,[9] and it is still the early days for the amount of wind, solar, and transmission that will need to be built out. Getting to global net-zero by 2050 will require six times more mineral inputs by 2040 than today, with demand for lithium growing by forty times and graphite, cobalt, and nickel by twenty to twenty-five times.[10] But new mines in countries from the United States to Serbia face opposition, permitting delays, and cancellation.

So the infrastructure needed to confront climate change is apparently stalled by democratic processes, while countries like China are building ultra-high-voltage transmission lines at breakneck speed. Climate advocates, despairing at the inability of our inadequate democratic processes to deliver, might wonder whether authoritarianism might be the price of the clean energy transition. Moreover, the path to building out renewables leads through supply chains in authoritarian regimes with human rights abuses—from rare earth production to lithium processing to the manufacturing of solar panels in Xinjiang. It may seem like we need to tolerate authoritarianism in order to meet the speed and scale of the transformation needed. Or, that we need to embrace it in order to push thorough the behavioral and structural changes needed. But this is a dangerous response to the challenges we face.

THE CLIMATE-AUTHORITARIANISM DOOM LOOP

Clean energy abundance is not going to be possible while there are authoritarian states that want to keep producing fossil fuels. More broadly, authoritarianism is both symptom and cause of climate change, not an answer to it. At the time of this writing, in 2022—events are moving so quickly that this feels like a required disclaimer—a food crisis and energy crisis are under way. A 2022 UN report on food security stated that 2.3 billion people were moderately or severely food insecure in 2021, and warned that inflation in consumer prices from the pandemic and measures put in place to contain it were worsening the picture.[11] Right now this is a price crisis. If the war continues and the weather is unfavorable, it may turn into a supply crisis.

Russia's invasion of Ukraine feels like the latest ratchet in a self-reinforcing system.

Call it a "climate-authoritarianism doom loop": Conflict impacts commodities → rising civil unrest and rising authoritarianism → more conflict. Commodity prices also increase the cost of building new green infrastructure, which becomes tougher to finance. Climate change gets worse, which in turns affects food production.

It's a simplistic model, and every reader should be asking, How robust are these links, especially with conflict and authoritarianism? Sure, this is all unfolding against the backdrop of rising authoritarianism: Trump, Orbán, Duterte, Modi, Bolsonaro, Maduro, Putin, Jinping, plus a handful of close elections—and all that was before the latest food and energy crises. But can't unrest lead to democratic revolution?

We can predict the correlations among conflict, commodities, authoritarianism, and more conflict based on observable dynamics from two recent socioeconomic disruptions. The first predictor is the 2007–2008 and 2010–2011 food price spikes. As background, the general consensus is that these earlier food price spikes had multiple drivers, including higher energy prices, biofuels demand, speculation, weather, and a depreciating dollar. The research on the relationship between these spikes and conflicts offers mixed results, but there have been findings that exogenous shocks (from outside a system) are linked with conflict, especially riots and protests.[12] Food prices were one driver in the Arab Spring uprisings, which did not deliver democratic revolution. The second predictor is the disruption generated from COVID-19, from supply chain disruptions to lockdown measures, which had economic and social ripples through the world. As in the food price spikes, the disruptions to the status quo did not translate into democratic revolution (nor did the sense of crisis result in meaningful investment in social systems or the energy transition). The emerging evidence is that the measures to respond to COVID-19 exacerbated inequality, benefited billionaires, and exacerbated repression in many countries, where the pandemic was used as an excuse for bans on protests, censorship, and police violence. In other words, the recent experience from the past decade and a half indicates that economic disruption and commodity shocks will more likely lead to repression than to democratic social transformation.

Climate change amplifies this vicious circle, becoming an element in these conflict–commodity–authoritarianism relationships. The science says the relationships between climate and conflict are complicated: climate has affected organized armed conflict, but other factors like low socioeconomic development and state weakness are more important.[13] In the near term, conflict will be driven by socioeconomic conditions and governance more than by climate change. But at higher levels of warming, the consequences of weather and climate extremes will increase vulnerability and increasingly affect violent intrastate conflict, according to the Intergovernmental Panel on Climate Change.[14] Where scientists tend to land is on something like, *climate changes amplify conflict risk*. And climate change has already reduced food and water security and will further undermine food security and nutrition.

So treat the climate-authoritarianism doom loop as a hypothesis. But anecdotally, and in the absence of better data, the doom loop is a possibility. And the media ecosystem described above will benefit from it, shape it, and further spur it on.

This widening gyre is not destiny. There are several policy and cultural offramps, including peace-building, debt reform, domestic support, international aid for people in countries dealing with food price shocks, and civil unrest leading to revolution toward a democratic system.

Yet the thing about this doom loop is that, as it gains momentum, it is harder to escape. It's hard to finance new infrastructure when steel mills are bombed out. It's hard to grow new food when the tractors are destroyed and there's no fertilizer because gas is too expensive for the factories to run, and meanwhile extreme weather is ravaging the crops. This may lead to interventions aimed at stabilizing the climate system as a means of holding the climate steady, such as solar geoengineering or sunlight reflection modification, which refers to blocking a fraction of incoming sunlight (solar radiation) to cool the planet. The most commonly modeled option for doing this posits a small fleet of aircraft that fly aerosol precursors into the upper atmosphere every day, creating tiny particles that float around the Earth for a year or so, blocking some sunlight. This strategy has only been simulated in computer models, so it is in very early stages of research.

Much has been written about whether solar geoengineering could be democratic, given that it would be a planetary intervention. On one side, some scholars argue that it is impossible to do it democratically,[15] and that research on the idea should be restricted because of this. On the other side, other scholars argue that it doesn't have a predefined trajectory, and could be governed via representative democracy, much like other things are.[16] These debates seem academic, because the twenty-first century will involve a lot of muddling through in imperfect, nonideal conditions. We should do research to better understand all the impacts that solar geoengineering might have, so that science can be better positioned to weigh in on decisions about it. It's possible that in a few years, there might be overwhelming popular demand for solar geoengineering, which both autocratic and democratically elected leaders would have a hard time resisting. A key thing to understand about solar geoengineering is that it is not a one-off intervention, but a multidecade or multicentury project. If the intervention is interrupted before greenhouse gas concentrations have declined—either though time or through carbon dioxide removal—then the warming that was being masked by the particles will happen in a rapid time frame, with devastating consequences. This is why solar geoengineering would need to be done in conjunction with reducing and eventually ceasing emissions, and within a governance structure that is capable of maintaining the intervention for a long time. It sounds like a terrible idea, but the relevant question is whether it is more or less terrible than the alternative scenario of climate change. Maybe solar geoengineering would buy some time to build social offramps from this doom loop, or maybe it would offer authoritarians a way to quell dissent by performing a strong response to climate change. We should think now about how to make better case outcomes more likely.

Our visions and tactics for creating the good future, the future we want, don't know how to deal with such a climate-authoritarianism doom loop. (Geoff Mann and Joel Wainwright observed as much in their book *Climate Leviathan*, an exception to this generalization—they describe "climate leviathan" as adaptation projects that will allow capitalist elites to stabilize their position amid ecological breakdown.[17]) The climate movement brings together people who care deeply about climate—but our theories of change about how to build the future we want are generally not engaging with the dynamics of the doom loop. We still tend to lack a clear analysis of what the failure of the Arab Spring means for on-the-street protest tactics, what Putin's grip on power means for strategies for ending fossil fuel production, what energy geopolitics and fossil fuel shortages mean on the ground for climate politics, or what high metal prices and the geopolitics of critical materials mean for the ability to build renewable energy at levels to supply sufficient reliable energy to everyone. These dimensions have not been central to the climate movement and its strategies, which typically focus on local, grassroots action in places and with people already sympathetic to climate action.

What happens to the climate movement when it does confront these things? What happens to it when it confronts these things in the context of this particular media ecology? "Within any climate justice movement that could possibly be effective or radical, we will encounter a deep desire for a planetary sovereign, one capable of the emergency measures to save life on Earth," Mann and Wainwright foresee. Some may see withdrawal as the answer to the problems of both the Internet and climate change, and veer into misanthropic variants of degrowth thought. We should consider another option.

REWIRING THE GLOBAL NERVOUS SYSTEM

So far, this essay has put forth several grim suppositions. First, that the Internet is a key obstacle to climate action and that we are wired together not by cables and packets but in a highly emotional entity currently directed by the motivations of capitalist firms. Second, the prevalent concept of energy democracy as community control of small-scale renewables is vulnerable to manipulation on these platforms, won't deliver the scale of climate action we need, and paradoxically may lead to consideration of authoritarian approaches. And, third, we may be in a climate-authoritarianism doom loop, for which the politics of the climate movement are ill-prepared.

What if this terrible thing that tech has built is also the very thing we need to get things moving in the right direction? What if we could reprogram this global limbic system, and redirect it toward empathy and building a shared vision of the future? Perhaps the world being networked together can offer the grounds for more global solidarity and tactics, making those TikToks of buildings washed into roiling rivers in Pakistan or images of withered rivers in France or China more than just triggers for anxiety and melancholy.

In broader society, climate change faces a lack of attention. But there now exists a system for focusing global attention and emotion. Rather than using it to stir up conflict and stoke fear, perhaps we could be using it to imagine better futures and focus on shared goals. At their best, the platforms offer glimmers of that—scrolling, one might catch real-world examples of green social housing, scientific discoveries, maps of the possible. To some extent, this presumes that platforms can be appropriated toward being tools, rather than systems that are set up to profit a few making *us* into tools. How? The answers will come only from collective thought, but here are two intuitions.

The first concerns dismantling the structure and building a new one; in many ways, this infrastructural challenge is akin to that of the energy transition, in terms of getting rid of an infrastructure that's harming us. Understanding what might be possible starts with being clear-eyed about how the platforms are actually working: how they condition our inner lives, our relationships, our institutions, our politics. Changing all of this isn't as crude as changing the algorithm or the business model, though both of those would be obvious first steps. Simply replacing a content feed that optimizes for outrage with one that optimizes for empathy or social support could also end up with creepy results. Content moderation is also merely a mitigative measure. It's not a matter of changing the emotional content that circulates on the platform, or even the algorithm that up-selects or down-selects that content. This function might help coalition-building for climate action, but it would still be antidemocratic in the sense that the algorithm is black-boxed and controlled by the few. (In the capitalist hell version of this, tech platforms would promote climate-friendly measures, surveil you as you adopt them—buying plant-based food or lowering your digital thermostat—and then take credit for changing your behavior and claim carbon offsets based on your good deeds, thus offsetting the carbon emissions of their own operations. Today's satire—or the logical endpoint of current trends?)

The real solutions are not in tweaking the signals in our global nervous system, but in changing the political economy of how the platforms direct where the signals move. The ultimate need is for public, collectively owned, and nonprofit platforms.

Momentum has been building to regulate the platforms, bolstered by events like the testimony to the US Congress from people like Frances Haugen, the Meta (Facebook) employee who explained how Instagram harms our children's health. In the United States, there is an explosion of debate about which approaches to take to regulation, whether by breaking up big platforms into smaller ones, or adopting a stricter privacy environment as in Europe, or implementing provisions for transparency, or enacting a tax on data, and so on. Reviewing these options is beyond our scope here, so just one contention: in terms of Internet and media reform, we should demand not just regulation or a breakup of these platforms, but public ownership. Works like Lizzie O'Shea's *Future Histories* or Ben Tarnoff's *Internet for the People* offer pointers in this direction. As Tarnoff writes, in creating different models for owning and organizing networks, democracy itself is at stake.[18] His book argues for deprivatization as a strategy, calling for creative experimentation in alternative collective networks that work in federations, using public institutions for procurement contracts that help the deprivatized sector grow, and tax breaks and loans and grants to support it. Advocating for public platforms should be a central demand of the climate and democracy movements.

There's a second intuition about how we rewire this global nervous system, which is more slippery, and has to do with how our culture processes emotion and its overall emotional intelligence. We may hope our response to climate change will be guided by rationality and science, but in order to get there, we have to leave room for emotions, too—for acknowledging past and ongoing harms from colonialism and environmental injustice, for example, and working through trauma, guilt, and more. When it comes to policy, this might involve enacting public processes around truth and reconciliation, funding mental health appropriately, encouraging genuine policy responses to loss and damage, or having national leaders serve as role models for how to speak openly about a range of emotions. Moreover, as with climate change, there is also a role for changing personal habits when it comes to rebuilding the Internet. Neither of these big dilemmas can hinge on behavioral change, but they can be aided by it—individual actions can strengthen political movements through shared experience. We need to move into new cultural norms about how to interact with each other, how to create and share content, and how to be in connection with a group. This means mundane personal actions like reducing time spent on social media, switching platforms, choosing civility, and so on.

The bottom line is that unless we understand that the structure of our media environment is a very real *climate action* problem, we will be stuck. There's a lot of talk about technological interventions into the climate system. The specter of solar geoengineering—the unceasing flights, the changed quality of the light, the unknown regional impacts—hangs over our halting efforts to confront climate change. It's mediagenic, the charismatic megafauna of the climate crisis space. But the sort of sociotechnical interventions we should be thinking about are also into the tech platforms that govern our social interactions, our emotional tenor, our political possibilities. Intervening in these spaces may preclude the need to intervene in our climate, and may disrupt the climate-authoritarianism doom loop. The story of the twenty-first century could also be the story of using technology to create forms of connectedness that support our common humanity. The setting could be our living rooms, schools, regenerating fields, and city streets. The characters could be the engineers of new online communities, the workers that maintain them, the educators and librarians that inhabit them, the groups of neighbors who have platforms not for complaining but for building new things—and you. The action is yet to be written.

7 COULD A GLOBAL CLIMATE REVOLUTION SAVE THE PLANET (AND DEMOCRACY)?

Frederick W. Mayer

As the planet careens toward the middle of the twenty-first century, two existential crises threaten humanity: the twin crises of climate and democracy. Alone, each presents challenges of enormous difficulty and profound stakes. Together, they are so daunting as to threaten paralysis. And they are intertwined: we cannot solve climate change without effective systems of democratic governance, and we cannot have effective democracy unless we make considerable strides in limiting those impacts of climate change that tend to undermine it.

It is almost impossible to overstate just how formidable the challenges are. In both its causes and its consequences, climate change is a truly global problem, and for that reason its solutions, more than those of any other issue to ever face the planet, must be truly global in scope. And, while meeting the challenge of climate change clearly has scientific, technological, and economic dimensions, it is fundamentally a political problem. Our utter failure to date to do what needs to be done is best seen as political failure on a global scale, a failure to translate the true interests of humanity into public policies at all levels of governance.

More precisely, the global political failure is an artifact of the global democracy crisis. It is tempting, but ultimately wrong, to believe that democracy is so cumbersome as to be incapable of making the hard choices necessary to deal with such a wicked problem. True, the characteristics of climate change make it a daunting problem for democracy, but pinning our hopes on benign autocracies is a chimera. In dealing with climate change, as with many other matters, democracy is the worst form of government—excepting all others. Alas, just when we need strong democratic institutions of governance, we see democratic backsliding around the globe driven by populist nationalism hostile to democratic values. From China to Russia to India to Hungary, and indeed in the United States—as measured by V-Dem, Freedom House, the Institute for Democracy and Electoral Assistance (IDEA), and *The Economist*—democracy is in decline.

Meeting the global climate crisis will require reversing that trend and developing more robust democratic institutions on all scales. However, the prescription that we need *first* to fix democracy before we can fix climate change is highly problematic as a practical matter. For one thing, climate change is itself a threat to democracy. Unchecked and un-adapted-to, the impact of climate change on migration, food insecurity, economic inequality, water scarcity, and climate-related conflict are all stressors on democratic governance.

Even more importantly, the defense of democracy requires a cause sufficiently motivating for publics to demand it. Abstractions about democracy are little match for the wave of nationalist populism now sweeping authoritarians into power. As we have seen throughout history, publics can be all too willing to trade democratic process for promises of salvation. Only when the disjunction between the interests of society and that of rulers becomes insufferably acute do societies typically arise to demand more voice. In this chapter, therefore, I argue that the climate crisis could be the sufficiently dramatic cause needed to catalyze a climate revolution. We just might be able to save democracy as we tackle the climate emergency.

Looking to a global climate revolution as the path forward for climate and democracy may seem unrealistic, perhaps even crazy. Indeed, that road faces formidable obstacles, but without significantly greater pressure from society, we cannot expect our existing systems of governance either to do what needs to be done on climate change, or to reverse democratic decline. And, as I will argue, a climate revolution is not so impossible. Already, the once abstract and distant impacts of climate change are becoming concrete and immediate. And this is just the beginning. As the climate worsens, the pressures will build. The open question is whether the societal forces they unleash can be harnessed to solve the climate crisis and, in so doing, save democracy, or those forces will condemn us down to some dystopian future?

THE CLIMATE CRISIS AS POLITICAL FAILURE

Often the climate crisis is characterized as market failure on a colossal scale, the consequences of billions upon billions of production and consumption decisions made without consideration of the costs they impose on the planet (i.e., in economist-speak, "negative externalities"). Climate change is the unintended by-product of the ubiquitous market economy, the dark side of the invisible hand. No one intended to destabilize the climate through the ways the market has structured transportation, agriculture and food distribution, industrial production, extractive industries, residential and commercial real estate, or all the other patterns of modern life—it just happened.

Market failure is not self-correcting. True, businesses around the world have taken some actions to limit their carbon footprint. Many have taken meaningful action to reduce their emissions or to ameliorate climate impacts. A large and growing percentage of electric power is generated with renewables, the electrification of transportation is accelerating, energy consumption per unit of GDP is declining, new technologies are on the horizon, and more. Yet private governance, alone, will not solve the problem for us.[1] Ultimately, we need public policies that establish the right incentives for private transactions.

For this reason and because we now know what is happening, why it is happening, and what must be done to address it, the ultimate failure is not that of the market, but rather of politics. If politics is defined as the process through which policy decisions are made and implemented by governments and other public institutions, the planetary political performance to date, a few bright spots excepted, is simply abysmal. At every level, from the local to the global, political institutions have failed to produce the policies we need if we are to avoid a climate catastrophe.

Two sets of policies are required. First, to mitigate the magnitude of climate change, a raft of policies needs to be more or less universally adopted to move to near-zero greenhouse gas emissions within an extremely short time frame. These could include market instruments such as a carbon tax, regulatory measures on carbon emissions and building standards, subsidies for renewable energy and conservation, protection of carbon sinks such as the Amazon and Congo rainforests, and many others. Second, to minimize the predictable societal impacts of climate change already baked into the system, significant policy commitments are needed to adapt to climate change, particularly to enable the world's most vulnerable communities to adapt. These include funding to limit flood impacts, enable agriculture to adapt to new climate patterns, address food shortages, respond to extreme weather events, and cope with large-scale migration.

In neither realm are current policies even close to adequate. On mitigation, we have yet to even begin to bend the upward trajectory of emissions. Since 1994, when the UN Framework Convention on Climate Change (UNFCCC) became the organizing vehicle for annual intergovernmental meetings, global CO_2 emissions have risen by 50 percent. Although the 2015 Paris Accords set a goal of limiting the global temperature increase to 1.5°C, the rhetorical national commitments made to date are inadequate to meet that goal, and there is little reason to believe that even they will be met. Indeed, globally, current policies seem likely to lead to somewhere between a 2°C and 3°C global temperature increase.

Three of the world's four largest emitters—China, India, and Russia—continue to trend in the wrong direction. China has pledged only to peak its greenhouse gas emissions by 2030 and to be carbon neutral by 2060, but in the meantime, it has blown past the United States as the world's largest emitter of greenhouse gasses (now fully a quarter of the world's greenhouse gas emissions). India's trajectory is equally troubling. Although in 2021 Prime Minister Modi announced new targets for 2030 and pledged carbon neutrality by 2070, its strategy for reaching those goals remains unclear. And Russia has now virtually seceded from international cooperation on all fronts. Even before its invasion of Ukraine, Russia's energy strategy focused on maximizing fossil fuel production and export. After its invasion of Ukraine, desperate for money, it has doubled down on that strategy.

The story in the United States is more positive, although still insufficient. After rejoining the Paris Accords when President Biden took office, the country pledged to reduce its net greenhouse gases 50 percent by 2030 relative to 2005 levels. The passage of major legislation in August 2022 makes it possible to come close to that goal. Still, an intransigent Republican opposition continues to block a more substantial policy response, and the US Supreme Court's ruling in *West Virginia v. Environmental Protection Agency* limits the possibilities for regulatory action.

On adaptation, the global record is equally abysmal, particularly with respect to aid for developing countries. The impacts of climate change are and will be felt most forcefully by developing countries, both by virtue of their geography and because of their relative lack of adaptive capacity, yet they have contributed very little to creating the problem. Naturally, they have argued that justice demands that they receive adaptation support. To date the demands of the developing world have fallen on deaf ears. Adaptation costs and likely adaptation financing needs in developing countries are five to ten times greater than current international public adaptation finance flows, and the gap seems likely to increase.[2]

Our political systems are clearly failing the test. Of course, it is hard to imagine another issue that presents anything close to as devilish a political challenge. Mitigation is something of a global public good: (almost) all nations would be better off if all cooperated in reducing greenhouse gas emissions, but every nation has the temptation to free ride on the efforts of others. This is a classic collective action problem (in game theoretic terms, it is a multiparty Prisoners' Dilemma or social trap game). Solving such dilemmas requires strong institutions that commit parties to mutually beneficial cooperation. But because the world lacks all but the weakest of international political institutions to facilitate agreements and enforce commitments, global cooperation is tenuous.

If that weren't enough of a challenge, the situation is highly asymmetric, with some nations standing to benefit more than others from mitigation (indeed a few northern nations might actually prefer climate change!) and some nations are able to mitigate in less costly ways than others. Differing perceptions of climate justice further complicate the problem, with developing countries arguing that rich developed nations that have historically contributed the lion's share of greenhouse gasses should therefore bear the brunt of mitigation efforts, while the developed world argues that the crucial question is ensuring that developing nations follow a sustainable path to economic growth. As a consequence, agreeing on what would constitute a fair commitment to mitigation is highly contentious.

The politics of adaptation are, if anything, even more difficult. Adaptation funding is essentially a transfer from wealthier nations to poorer and more vulnerable ones. It is a form of foreign aid, and like all foreign aid, it is much harder to convince donor countries that the benefits of aid outweigh the costs unless there is some quid pro quo.

Finally, and crucially, the international politics of climate change is a two-level game: stances in international negotiations and the national policies needed to make and keep international commitments are the artifacts of domestic politics.[3] And political failure is rife in the domestic politics of virtually all nations. Typically, with respect to mitigation, the voices of environmental interests are at a considerable disadvantage compared to entrenched energy producing and consuming interests. Couple that with disinformation, the immediacy of costs of acting and the long-delayed benefits, and simple inertia, taking policy action on climate change becomes a very tough sell. The politics of funding for adaptation is even more difficult. Just as foreign development assistance is chronically underfunded, so too there is little appetite in wealthier nations for transferring funds to help poorer nations adapt.

In sum, across the globe there is a massive political failure to produce the policies that are needed.

WHY DEMOCRACY?

Given that the characteristics of climate change are so politically challenging, might we be asking too much of democracy? Indeed, it is tempting to imagine that the solution to the climate crisis is to centralize power in the hands of enlightened autocrats who could override the dysfunction of politics. That is always the appeal of autocracy. Alas, there is no easy out here. We do not have and will not have (and would not want) a global Leviathan. For better or worse, we are in a world of weak international institutions, and therefore dependent on the collective action of more-or-less sovereign nation-states.

But what about national autocracies? Because they do not face the same domestic political pressures as more democratic states, might they be better positioned to do the hard things that need doing?

Many, in asking these questions, have China in mind, and there are ways in which China's government appears to do be doing better than Western democracies, notably in nurturing its green energy capacities. But the record of environmental performance by autocratic regimes, and indeed on closer inspection of China itself, is not encouraging. Many autocratic regimes are essentially petrostates, among them Russia, Saudi Arabia, Iraq, Iran, the UAE, Kuwait, Venezuela, Qatar, Nigeria, and Kazakhstan, all deeply dependent on fossil fuel production and deeply compromised. China, for all its advances on renewables, remains highly reliant on coal, with one of the highest rates of CO_2 emissions per unit of GDP in the world.

Elsewhere where the trend has been away from democracy—notably Brazil, India, Russia, and indeed the United States—less democracy has meant less attention to climate change. Indeed, it is generally the more democratic countries that have the most progressive climate policies. It is also worth noting the high correlation between autocracy and extreme nationalism, which tends to make autocracies less cooperative in the international arena. No one would argue that the autocracies are leading the charge in the annual climate global climate conferences, for example.

That autocracies are less likely to reduce their carbon footprint to cooperate in a global carbon regime should not be surprising. The heart of the matter is that they are generally less responsive to the interests of environmentally concerned citizens. Autocracies have far less well developed civil societies, with fewer opportunities for political mobilization on all matters, climate included. Moreover, far from being more farsighted, autocrats' preoccupation with maintenance of power often leads to the prioritization of immediate rather than long-term rewards. In short, autocracies are less motivated to address the climate crisis because they have a democratic deficit.

Unfortunately for the planet, democracy is in decline around the world. There is clear democratic erosion in all five of the world's most populous countries: Russia, China, India, Brazil, and, yes, the United States. According to V-DEM, fully 70 percent of the world's population now live in some form of autocracy. Freedom House reports that freedom around the world has declined for sixteen straight years. *The Economist*'s democracy index is at its lowest measure since it was launched. And as IDEA's annual *Global State of Democracy* report puts it:

> The world is becoming more authoritarian as non-democratic regimes become even more brazen in their repression and many democratic governments suffer from backsliding by adopting their tactics of restricting free speech and weakening the rule of law.[4]

Anything close to a full treatment of the causes of the global democratic decline is beyond the scope of this essay. They surely include the role of social media, more skillful use of propaganda by authoritarians, the discouraging example of the United States, and more, but among the likely culprits are the impacts of climate change. The pathways through which climate change is stressing democracy include "more frequent and severe natural disasters, increasing food insecurity, economic decline and financial instability, climate injustice, and climate-related conflicts and migration."[5] Each of these climate impacts is destabilizing for democracy and provides fertile grounds for the promises of the strongman. In no small measure, therefore, saving democracy will require solving climate change.

THE COMING CLIMATE REVOLUTION

So we face a conundrum: we need democracy to solve climate change and we need to solve climate change to save democracy. It is hard to know where to intervene. Indeed, it is hard not to fear that we are doomed to a death spiral, with each crisis reinforcing the other. And yet act we must.

One simple formulation is that we first need to solve democracy to save the climate. That is fine as far as it goes. We should be doing all we can to address the crisis of democracy: combating misinformation, strengthening voter rights, supporting efforts to bridge divides, facilitating civil discourse, and more. Any progress on arresting or reversing democratic erosion would not only be desirable on its own terms but would also likely have salutary effects on the planet's climate.

But there is good reason to worry that a fight for democracy is far too abstract a concept to generate the political energy it will take to turn the tide. As we have seen throughout history, and certainly see around the world today, publics can be all too willing to abandon democracy if promised that there is an easier means to ends they desire. And the more stress a society is under, the greater the siren call of the strong leader who will restore the imagined better past. It is an immensely seductive narrative. As a recent Pew survey of global public opinion summarized, "People like democracy, but their commitment is often not very strong."[6]

What, then, can be done?

We cannot rely on the current political system to fix itself. At the international level, the entire liberal order is under great stress. More than a decade ago, Robert Keohane and David Victor argued that the fragmented nature of international regimes coupled with domestic political constraints made it unlikely that the "climate regime complex" would be able to meet the climate challenge.[7] Matters are not much better today. In this era of heightened nationalism, there is little likelihood that international processes, alone, will lead to democratic and effective global institutions capable of generating the policy responses necessary to combat climate change. Similarly, democratic erosion within national governments is not self-correcting. China's neo-authoritarianism is frighteningly entrenched. There are few currently evident prospects in other authoritarian states of democratic revival. India is on a very worrisome trajectory. And in the United States, presented with clear evidence of what amounts to a coup attempt by a leader with authoritarian instincts, it is unclear whether the polity can be roused to defend democracy. The bottom line is that without some fundamental change in the political context, it is hard to see where the impetus will come from to fix democracy.

I argue, therefore, that political change of the necessary magnitude can be catalyzed only by transnational societal pressure sufficient to alter the political (and economic) calculus of governments (and businesses), that is, a global social movement. This is a very tall order indeed, but without it, the inexorable logic of the market and the crude calculus of politics cannot be countered. Only an issue of sufficient urgency, in which publics are truly aroused by the failure of their political institutions, can generate the political energy to demand democratic change. And that issue is the climate crisis.

To be sure, a global climate social movement faces formidable odds. Collective action on such an extraordinary scale, across so many cultures and across all manner of differences, even if only loosely coordinated, is an extraordinary challenge. It requires, first, establishing a shared understanding of the stakes involved and of humanity's common interests. Moreover, because movement requires overcoming the temptation to free ride on the actions of others, it requires that acting be its own reward independent of the impact of that action on the outcome. Whether voting, joining, giving, or advocating, no one person's act will have any appreciable impact on the outcome.

To date, we have seen fragmented international activism, impressive but not yet sufficient to override the political logics that currently hold sway. But the conditions for movement are becoming more favorable. Publics are slowly coming to the realization that the crisis is upon us. In the United States, where climate skepticism has held policy back, the 2021 Yale survey found that 72 percent of Americans now believe that climate change is happening, and 57 percent believe that it is caused by humans.[8] Surveys by Pew, ABC/Washington Post, and others find comparable figures. Around the world, opinion varies by country, but a 2022 Yale survey found that a clear majority in every country now believes that climate change is happening[9] and a 2021 UNDP survey found that, on average, 64 percent see climate change as an "emergency."[10] Generational change will help as well; younger people are both better informed and more concerned.

The gap between opinion and reality will surely continue to narrow. We are just beginning to feel the impacts of climate change. Over the next decade, as extreme heat, droughts, violent storms, rising sea levels, and all the other impacts of climate change become more common, the contours of the crisis will become ever more apparent. As will the consequences: food shortages, deaths from heat, large-scale migration, water shortages, and climate conflict.

Beyond changing attitudes, if one looks closely, there is climate action bubbling up in many quarters of the world. Civil society organizations focused on climate continue to proliferate. Businesses are greening their operations and looking for ways to make sustainability consistent with their business strategies. Impact investing continues to grow. Consumers are becoming more climate conscious. There is a buzz about regenerative agriculture. And these efforts are not isolated. Activists in all these realms share a common understanding of the urgency of the moment and a common determination to press for better results from politics. Furthermore, they are increasingly connected to each other across the planet.

There are also some positive signs that climate concerns are translating into government action. Cities around the world are taking the lead; many have sustainability plans. In the United States, California is banning gasoline-powered cars and Congress, paralyzed as it has been by America's political dysfunction, managed in 2022 to narrowly pass legislation that will give a major boost to renewable energy in the coming years. Many countries in Western Europe have gone considerably further. These are evidence that climate concern can be translated into public policy.

At the same time, there is high and growing dissatisfaction with democratic governance around the world. Too often, of course, that dissatisfaction is being exploited by antidemocratic forces, but it could be harnessed to demand more democratic and responsive governments. For example, the belief that the state is run for the benefit of everyone in society has decreased significantly over time. A 2021 Pew Research survey of global public opinion found that "for many, democracy is not delivering; people like democracy, but their commitment to it is often not very strong; political and social divisions are amplifying the challenges of contemporary democracy; and people want a stronger public voice in politics and policymaking."[11]

All of this is still some distance from a global climate revolution, but the seeds are there. As what was once distant and abstract manifests in the form of heat waves, droughts, floods, forest fires, crop failures, and more, the pressure for action will build. Advocacy groups will find it easier to recruit members. Consumers will demand more from businesses. Citizens will ask more of their governments. And generational change is bringing a more receptive public into the arena.

To be sure, a climate revolution that succeeds in addressing both climate change and democratic backsliding is far from preordained. Dark scenarios are all too possible. As the pain of the climate crisis becomes more extreme, predictably, discontent with the failure of our politics to address it will grow. Will that anger lead to more democracy or to its dissolution? Both are possible. Under stress, democracies have often collapsed but that is not inevitable, and dissatisfaction could be turned toward improving democracy.

Converting growing dissatisfaction into a democratic climate movement will require leadership. There is work to be done in many arenas. In national politics, public leaders need to forge alliances and build political coalitions capable of overcoming the forces that have blocked climate action. This will be harder in some countries than in others—the politics of China and Russia seem, for the moment, impervious to outside influence—but success in places with both robust climate activism and still-functioning democratic institutions, most importantly the United States, would encourage action elsewhere. In the international arena, a coalition of democratic and climate-conscious nations could take the lead both within the UNFCCC process and in developing new governance institutions more responsive to societal demands and better able to facilitate international cooperation. Within nations much of the work will need to come from societal leaders who can channel the coming public discontent in to constructive demands.

To animate, coordinate, and sustain anything like a global movement sufficient to motivate democratic action on climate change will require what all social movements require: a powerful narrative that enlists humanity in its cause. As I have argued elsewhere, a shared narrative is almost always at the heart of large-scale collective action.[12] Narrative is humanity's "go-to" tool for constructing common understandings of our circumstance, for seeing dangers over the horizon, for creating interests not just in ourselves but in the fate of others, for enlisting each in a collective cause, and for making acting in pursuit of that cause itself meaningful, a necessary expression of our character in the story of our lives. A central task of leaders, therefore, is to tell the story, to invite publics to imagine themselves as actors in an epic narrative, in which humanity faced its greatest peril, and in which it exercised its collective voice and rose up to demand more responsive policies of governments, and in so doing rescued the planet from climate disaster and restored confidence in the possibilities of democracy.

Leaders in politics, in business, in the media, in the arts, and in the academy, whatever other roles they may play, can all be storytellers, connecting the dots, crystalizing the stakes, and lighting the way out of the dark corner we have painted ourselves into. That leadership will be critical as the climate pressures mount, and demagogues seek to distract publics from the real task at hand by vilifying migrants, whipping up nationalistic fervor, and promising a return to an imagined glorious past—the usual narratives of autocrats.

Calling for a global climate revolution as the solution to both the climate and democracy crises may seem like a long shot. Yet there is little prospect that the current political system will respond with the urgency that the crisis demands. When all other roads are blocked, a long shot may be our best. And it is not so wild a dream.

III POLICY AND LAW

8 BREAKING POLICY GRIDLOCKS

William S. Becker

It has been 160 years since the first scientist confirmed that carbon dioxide (CO_2) pollution was changing the Earth's temperature. It was not until 1992 that the world's 197 nations and states agreed to do something about it. Now an entire generation has passed, so why are scientists still saying we are in danger of worldwide social collapse or even human extinction?[1]

The simple answer is that no nation has embarked on the "rapid, far-reaching and unprecedented change in all aspects of society" that the Intergovernmental Panel on Climate Change (IPCC) says is necessary to keep the Earth livable. Under a breakthrough agreement approved unanimously in Paris, nations are developing plans to reduce their climate-changing greenhouse gas emissions, but their plans fall short so far. Greenhouse gas emissions continue adding to the blanket of pollution covering and warming the Earth. The international community is effectively gridlocked because doing too little is as bad as doing nothing when a threat is existential.

People are understandably prone to apocalypse fatigue when they hear repeated warnings like this, but that doesn't change reality. Unless we break gridlock, it has the potential to break us, and even to break the civilization humanity has been building for the past twelve thousand years.

INSUFFICIENT AMBITION

After nearly a quarter century of talking, nations agreed, in the Paris Accords, to limit global warming to 1.5°C above preindustrial temperatures or, failing that, well below 2°C. However, cautious about protecting their sovereignty, nations agreed to submit *voluntary* rather than binding plans to reduce their greenhouse gas pollution. Although global warming is approaching the point where it will be irreversible, the plans are still insufficient to achieve the Paris goals. Why?

The political and economic barriers in each nation are different. But to give a sense of the causes of gridlock, this essay explores the situation in the world's second-largest source of CO_2, the United States. The country has the intellectual and fiscal resources to achieve a net-zero-carbon economy by midcentury, the objective widely adopted by nations, corporations, and communities worldwide. Nevertheless, 80 percent of America's energy comes from fossil fuels, whose emissions are the principal cause of climate change. Analysts predict that without significant changes in energy policies, fossil fuels will still be the United States' biggest energy source in 2050—the year nations have agreed to achieve the decarbonization of the world economy.

CAN DEMOCRACIES HANDLE IT?

In his book *Can Democracy Handle Climate Change?* Daniel Fiorino argues that democracies have attributes that allow them to handle climate crises better than authoritarian regimes, including "an open flow of information and opinions through freedoms of speech, press, and assembly" and the ability to choose leaders.

That might be true for healthy democracies, but democracies worldwide are undergoing a "precipitous decline," according to a 2022 evaluation by Britain's Economist Intelligence Unit (EIU). Only 6.4 percent of the world's people live in full democracies, while more than half live under authoritarian or hybrid regimes. The EIU ranked the United States a "flawed democracy" for the sixth year in a row.

Deep political divisions in the United States, in Congress, and in US society are primarily responsible for its low rating. US democracy, once considered a model for the world, is itself in crisis. Domestic extremists are weaponizing the rights guaranteed by the US Constitution. They use the right of free speech to spread disinformation and conspiracy theories, claim freedom of assembly to gather for violence, and use the Second Amendment to arm antigovernment militias with military-grade weapons.

More to the point, the fossil energy industry virtually controls national energy policy, as it has for more than a century. Years ago, Republicans in Congress conspired with fossil-energy lobbyists to make climate change a wedge issue by promoting old myths that clean energy will kill more jobs than it creates, raise energy prices, and leave the US short of the energy it needs. But with the weather becoming increasingly destructive and scientists' dire predictions coming true, public attitudes are shifting in the United States and elsewhere. People are getting impatient. In 2022, a Politico poll found that

> Adults across the United States and globally have damning opinions about the performance of their political leaders when it comes to climate change and say they are noticing an escalation in extreme weather events and natural disasters. . . . [The poll] reveals frustrations from citizens that they are being left to take climate action on their own when they believe governments and the companies with the most resources (which also tend to bear the most responsibility for carbon emissions) should shoulder the burden.[2]

That brings us to the principal barrier to more aggressive climate action in the United States: the continuing political power of the fossil energy sector.

STRANDED INVESTMENTS?

The sector and its investors are gambling they will be part of the decarbonized world. The International Energy Agency (IEA) says the path to a net-zero carbon economy "calls for nothing less than a complete transformation of how we produce, transport, and consume energy." There are no new oil and gas fields or coal mines on the path.[3]

Research published in 2021 concluded most recoverable fossil fuel reserves—60 percent of oil and fossil methane, along with 90 percent of coal—must remain in the ground for a 50 percent chance of achieving Paris Accords goals.[4] Nevertheless, the world's largest banks financed nearly

$4.6 trillion in fossil energy projects after adoption of the accords. The IEA expected the industry to make $200 billion in capital investments in 2022 and to double its net earnings to a record $4 trillion.[5]

In the United States, oil and gas companies had more than four thousand miles of new pipelines under construction in 2022 with another 17,800 miles planned. Oil and gas producers held drilling leases on forty million acres of public onshore and offshore lands in 2022, but they want more. In August that year, after thirty years of inaction on climate change, the US Congress finally approved a $370 billion investment in clean energy. But it was able to pass the bill only after agreeing to allow oil and gas production on another sixty-two million acres of federal property.

Meantime, global warming is making bad weather catastrophic. Some $1.75 trillion of coastal real estate is at risk of sea level rise, the US West is running out of water because of an unprecedented drought, history's most intense wildfires are burning down entire communities, and daily temperatures hot enough to kill are common. Natural disasters displaced nearly ten million Americans from 2008 through 2020. Yet, still attracted by the oceans' magnetism, more Americans are moving into rather than out of harm's way.

WHAT TO DO

"In many ways, gridlock is endemic to our national politics, the natural consequence of separated institutions sharing and competing for power," according to Sarah Binder, a senior fellow at Brookings. "If a government that 'governs least governs best,' then policy stability should be applauded, not derided as gridlock."

But we should hold the applause for governments that govern least regarding climate change, because their people will suffer most. In the United States, state and local governments have filled the leadership void left by the federal government, but that cannot continue. The country cannot accomplish an energy transition deep enough or rapidly enough with the federal government on the sidelines. So, here are some of the things the three branches of US national government could do to break their long-standing gridlock on climate action.

Nationalize Oil and Gas

The global economy will achieve net-zero carbon when its CO_2 emissions are in balance with what nature and technology can absorb or prevent. Several big US oil producers have signed up to the net-zero goal, but it has high greenwashing potential. Polluters can keep emitting CO_2 by purchasing "carbon credits" that pay for clean energy and carbon sequestration projects elsewhere. There are no agreed-upon standards to prevent double-counting (two or more companies buying the same credits) or buying into projects that would have happened anyway. In addition, companies that promise to cut their emissions often are talking about their supply chains, not their own operations. Yet McKinsey notes "supply chains have the greatest room for improvement to meet sustainability goals . . . Supply chain impacts account for more than 80% of greenhouse gas emissions and more than 90% on air, land, water, biodiversity and geological resources."

In short, even though the transition to clean energy has been called the most significant market opportunity in history, the major oil and gas companies have given us no reason to think they are willing to leave their recoverable reserves in the ground. It's apparent that the US transition to clean energy will not happen, or will not happen rapidly enough, unless the government nationalizes oil and gas production.

The government has nationalized strategically many times when it felt it was necessary, most recently to rescue the savings and loan industry in 1989 and to bail out banks, investment houses, insurers, and the US auto industry in 2008.[6] Political leaders considered these industries "too big to fail." Now the Earth and civilization itself are too big to fail.

One way to do this, economist Robert Pollin suggests, would be for the government to buy controlling ownership of at least ExxonMobil, Chevron, and ConocoPhillips, the three dominant oil and gas corporations. Pollin estimates the price would be about $420 billion (based on stock prices in April 2022), a sum only one-tenth as large as the $4 trillion the Federal Reserve authorized in 2020 to sustain the corporate sector during the COVID-19 pandemic.[7]

It does no good to shame the industry into doing what's right. Its mission is to make profits for shareholders. It's the government's mission to do what's best for humanity.

Balance the Courts

Frustrated at government inaction, activists and global-warming victims in several countries began turning to the courts in the mid-1980s. The *Guardian*, a British newspaper, reports that climate-related lawsuits worldwide have doubled since 2015.[8] By 2022, there were more than two thousand. The IPCC calls litigation necessary to force reluctant fossil fuel companies and governments to act against global warming.

The lawsuits' goals range from stopping the construction of coal-fired power plants to penalizing fossil fuel companies for misrepresenting climate science. Other cases are trying to establish innovative precedents like the rights of nature, and laws against "ecocide."[9] In addition, children and young adults have gone to court in several countries to establish that public officials have a fiduciary duty to protect the atmosphere as a public trust asset for current and future generations.

But climate litigation faces an uncertain future in the United States when cases reach its highest legal authority, the US Supreme Court. Six of the court's nine justices are conservative Republican appointees who have shown a bias toward the fossil-energy sector. In June 2022, the court sided with two coal companies and two coal-producing states to invalidate the government's most potent tool against CO_2 pollution, a national regulation limiting carbon emissions from coal-fired power plants.

Several of the court's recent decisions revealed that it leans far to the right, leading to calls for President Biden to "stack" the court—in other words, increase the number of justices to create better ideological balance at the top of the judicial system. That would set off a political firestorm. But law professor Dawn Johnsen, a former acting US assistant attorney general, points out that several of the current conservative justices were appointed under questionable circumstances—for example, by presidents who did not win the popular vote and, in one case, in a Republican ploy that denied Democratic president Barack Obama his right to appoint a justice.

Because federal judges and justices serve lifetime appointments, the Supreme Court's far-right tilt will continue for decades unless a president appoints additional justices for ideological balance. Congress should authorize the president to stack the court as soon as possible.

Allow Direct Democracy

The United States is a representative democracy where voters choose the public officials who make laws. Unfortunately, representative government is not always responsive government. When representatives are unable or unwilling to resolve issues of great public concern, governments lose credibility among their people, and democracies can slip into autocracies.

Elected officials might be more inclined to resolve impasses if they knew voters could take matters into their own hands. Twenty-three US states allow direct democracy, wherein citizens create, repeal, or amend laws their legislatures have approved. Since the beginning of the twentieth century, states have permitted ballot initiatives more than two thousand times, although only 850 resulted in legislation.

A similar process at the federal level would require a constitutional amendment approved by two-thirds of Congress and three-fourths of states. Without enormous public pressure, it's doubtful lawmakers would give up some of their power. But if citizens are as frustrated as the Politico poll indicates, civic organizations could petition Congress in overwhelming numbers to pass the amendment.

In the meantime, President Biden could follow the lead of other countries that have engaged in direct government, including France, where President Emmanuel Macron appointed 150 people to the Citizen's Convention on Climate and charged them with recommending how the nation could meet its greenhouse gas reduction goal. The exercise taught several lessons for direct democracy, most importantly to avoid overpromising whether and how governments use nonbinding results. Although Macron promised to pass the Convention's "unfiltered" proposals through to the French parliament, his eventual climate bill contained only 40 percent of the citizens 149 ideas. Nevertheless, Macron's government appointed a panel of randomly selected citizens to advise it on the country's COVID vaccination process.[10]

Depoliticize Science

During the administration of President George W. Bush, a whistleblower disclosed that the White House was censuring and editing the research of federal climate scientists to make their findings seem less confident. The

Trump administration was not so subtle; it proudly dismantled "core elements of the federal climate science apparatus."

Congress should require the executive branch to create a firewall between politics and science. It should forbid elected officials and political appointees from editing and censoring government research, limiting its release to the public, or otherwise hampering the government's science work.

End the Influence of Money

In 1974, Congress established a system for voluntary public funding of political campaigns. However, candidates seldom use the system because its funds are considered insufficient for a modern campaign.

In 2010, the US Supreme Court opened the floodgates by ruling government can't limit campaign contributions by corporations and wealthy individuals because money is a form of free speech. In the eight years following the decision, oil and gas industry donations to campaign committees climbed from $35 million to $84 million.

"The justices who voted with the majority assumed that independent spending cannot be corrupt and that the spending would be transparent, but both assumptions have proven to be incorrect," the Brennan Center for Justice reports.[11]

The influence of special-interest money may explain why members of Congress vote against the expressed preferences of their constituents. Professors Martin Gilens of Princeton University and Benjamin Page of Northwestern University analyzed two thousand public opinion surveys and compared them to Congress's actions. They concluded, "The preferences of the average American appear to have only a minuscule, near-zero, statistically non-significant impact on public policy" while campaign contributions from economic elites "still carry major influence."[12]

The Supreme Court erred in its naïve assumption that campaign contributions don't influence votes or even give that appearance. Watchdog groups should document cases in which special interests have effectively bribed members of Congress with campaign cash, then force the court to revisit the issue.

Restore Majority Rule

The filibuster is the practice of debating a bill so long that Congress is compelled to move on to other business. It is the procedure used by the minority party in the US Senate to prevent the majority party from passing legislation. It is the rule most responsible for congressional gridlock.

In his writings about America, Alexis de Tocqueville warned about the "tyranny of the majority," but the filibuster is an example of overzealous efforts to protect democracy. It results in the tyranny of the minority. Today, marathon speeches aren't necessary; any senator can stop action simply by declaring a filibuster. A difficult-to-achieve supermajority of sixty senators is required to resume regular business, so filibustered bills usually die.

There have been more than two thousand filibusters in the Senate since 1917, about half in the last dozen years as partisan politics has become rigidly polarized. Both parties use the rule. They keep it in case they end up in the minority. The Senate should repeal it and reinstate majority rule.

Forbid Conflicts of Interest

Members of Congress from both political parties have substantial holdings in companies and industries the lawmakers could help in the course of their public duties. A *New York Times* investigation found that, in recent years, nearly one in five members of Congress bought stocks related to their committee memberships.[13] The *Harvard Business Review* calls this "a growing conflict-of-interest problem in Congress."[14] The *Review*'s research showed members of Congress earned "abnormally higher returns" than other investors.

A prime example occurred as Congress considered its recent big climate bill. Democrats needed all of their senators to pass it. However, two held out for concessions, including Senator Joe Manchin, a Democrat who represents the coal state of West Virginia and chairs the Senate committee with jurisdiction over energy issues.

The watchdog organization Open Secrets reported that, although Manchin will not stand for reelection until 2024, he received more money from the oil and gas industry than any other member of Congress in the first half of 2022. The *Washington Post* found that Manchin received nearly a half million dollars in interest, dividends, and other income in 2020 from a family coal business in which he had shares worth between $1 million and $5 million. Manchin leveraged his power as the bill's decisive vote to win concessions for fossil fuels. (The second holdout, Senator Kyrsten Sinema of Arizona, also voted for the bill after sponsors incorporated her suggestions.)[15]

The Senate should amend its ethics rules to require that senators recuse themselves from votes when they have a significant financial interest in the outcome.

Fight the Fear and Outrage Machine

The attitudes that divide Americans today are fed by a loosely organized coalition of people who use social and conventional media to peddle fear and outrage. They demonize the political left, undermine confidence in democracy, malign public figures, promote racism, and openly threaten insurrection and civil war. Facts and truths are immaterial. Politicians and media stars get rich selling fear because so many citizens are willing to buy it.

The news media in the United States enjoy a special status: the First Amendment to the Constitution expressly guarantees freedom of the press. However, there is a long-standing tension between that freedom and the broadcast media's obligation to use it fairly. In 1949, the Federal Communications Commission (FCC) instituted the Fairness Doctrine, which required broadcast stations to present both sides of controversial public issues. A network could lose its broadcast license if it didn't comply. But as cable and satellite stations proliferated, the policy fell into disuse, and the FCC formally repealed it in 1985.

Fortunately, the FCC still has a regulation against deliberate false or slanted news, an abuse it calls "a most heinous act against the public interest." Unfortunately, a study in 2001 found the FCC rarely enforced the rule.[16] When it did, the punishment was usually no more than a written scolding.

Protected speech in the United States is not absolute. The Constitution's guarantee doesn't include incitement, defamation, fraud, obscenity, child pornography, fighting words, or threats. Congress should pass a law clarifying that speech fostering hatred, violence, insurrection, or racism falls within the unprotected category under threats, fighting words, and incitement.

Enforce "Settled" Environmental Policy

All three branches of the US government are responsible for reasonable policy consistency to maintain stability in society and the economy. When the US Supreme Court repeatedly upholds the decision of a past court, it becomes "settled law." We might say the same for an act of Congress meant to withstand the test of time.

The Ninety-First Congress established one of America's most important "settled policies" in the National Environment Policy Act of 1969 (NEPA). The statute begins with the statement that the "continuing policy" of the federal government is to

> use all practicable means and measures, including financial and technical assistance, in a manner calculated to foster and promote the general welfare, to create and maintain conditions under which man and nature can exist in productive harmony, and fulfill the social, economic, and other requirements of present and future generations of Americans. [17]

This policy has not changed, but circumstances have. In 1969, the connection between global climate change and fossil fuels was not a prominent issue. Now, "settled science" concludes that fossil fuel pollution is a permanent nationwide threat. Without mitigation, it will undermine the rights to life, liberty, and the pursuit of happiness of US citizens.

NEPA's policy statement strongly suggests that Congress must stop subsidizing fossil fuels. If climate change is allowed to proceed, man and nature clearly won't exist in productive harmony. The newly captured tax revenues would be much better spent on clean energy.

Create a Common Cause

Insurrection is not the greatest threat to democracy. Voter suppression isn't either. The greatest danger is complacency—the failure of the silent majority of US citizens to rise to the defense of democracy when it is under attack, as it is now. The Internet allows radicalism to spread much more quickly than it used to, especially when radicals have the national stage to themselves.

US citizens have a history of rallying and working together against common enemies. Now the enemy is greenhouse gas pollution and other causes of environmental degradation. The president of the United States should rally the citizenry to repair the ecological damages and restore the ecosystem services we've lost.

How much are the services worth? In 2014, researchers concluded that ecosystem services worldwide contribute more than twice as much value to human well-being as global GDP.[18] They estimated land-use changes were destroying benefits worth as much as $20 trillion annually. The World Bank says protecting critical ecosystems could avoid global economic losses of nearly $2.7 trillion annually by 2030. Similar data about ecosystems in the United States are harder to come by, but the benefits would be enormous.

Here are three suggestions on how to proceed. First, the White House creates a user-friendly searchable database of the scores of scattered federal partnership and grant programs that protect, restore, and sustain ecosystems and their services. Since these programs already exist, the campaign would not require additional federal spending.

Second, the administration continues restoring the many environmental standards rescinded by the Trump administration. That would revitalize America's "restoration economy," a sector that employed more than 220,000 Americans and produced $24.5 billion of annual economic output in 2015.[19] In addition to creating jobs and recapturing free eco-services, the campaign would advance environmental justice. Research shows that the loss of ecosystems has disproportionately affected lower-income and urban populations.

Third, President Biden uses his bully pulpit to launch a national Restore America campaign in partnership with organizations such as AmeriCorps, the Boy Scouts, the National Association of Cities, and Republican organizations such as ConservAmerica, Conservatives for Responsible Stewardship, and Republicans for Environmental Action. Restoring productive harmony with nature is the real way to make America and the world great again.

Employ Systems Thinking

NEPA contains another key passage about the proper approach to public policy. It says all government agencies should

> utilize a systematic, interdisciplinary approach which will insure the integrated use of the natural and social sciences and the environmental design arts in planning and in decision-making which may have an impact on man's environment.[20]

The key is the integration of relevant disciplines and a systematic approach to policy. Systems thinking is critical if we want twenty-first-century energy policies to be free of uncounted costs and unintended consequences. Unfortunately, US policy-making suffers from tunnel vision today. Every Btu of energy we consume is the product of a long value chain whose links stretch from the mine or wellhead to the car, factory, or power plant, and finally to the atmosphere, air quality, public health, and climate stability. Unfortunately, the cost-benefit analyses in energy policy often focus only on the moment fossil fuels combust.

A prominent example is a technology that many organizations and experts consider essential for achieving a zero-carbon world. Carbon capture and sequestration (CCS) separates CO_2 from other flue gases at power plants and industrial operations, then compresses it for transportation to a site where it is injected into the ground for permanent storage. CCS advocates hope it could reduce the world's greenhouse gas emissions 14 percent by 2050.

Unfortunately, CCS is meant to trap CO_2 immediately after combustion of a fossil fuel, but not the significant emissions and other environmental impacts along the fuel's entire value chain. Coal, oil, and natural gas affect the environment and produce greenhouse gases when they are taken from the ground, processed, and transported to the point of use. If CO_2 is captured at a power plant, it is compressed and moved to sites where it's injected into the ground for permanent storage.

Besides the environmental disruption and greenhouse gases produced by this process, CCS requires larger power plants, more cooling water, and parasitic energy to compress the gas. An estimated 68,000 miles of new pipelines would be necessary to transport captured CO_2 to injection sites in the United States. Most CCS installations today try to become cost-effective by selling their extracted CO_2 to oil companies, which inject it into wells to "enhance" oil production. Oil's value chain also produces CO_2 emissions that negate part of CCS's climate benefits.

Because of its many additional expenses, CCS would raise consumer electric bills substantially higher than the cost of power from solar, wind, geothermal, and hydroelectric technologies. The price of CCS power is so high that it still is not ready for widespread commercial use despite decades and tens of billions of dollars in government research.

As researchers and policy-makers consider how to produce energy in a decarbonized economy, their analyses should consider the entire value chain of each option, and they should make their analyses transparent. Anything else invites analytic sleight of hand.

Anticipate the Sovereignty Barrier

Nations' worries about sovereignty were a central factor in the long delay between 1992 when countries signed the United Nations Framework Convention on Climate Change and 2015 when they approved the Paris Accords. Nations accepted the Paris Accords largely because they call for governments to submit *voluntary* plans to combat climate change. There are no penalties for failure. There may come a point when some nations do not keep their commitments, and enforcement becomes necessary to keep the Paris Accords from dissolving.

There are already examples of unkept commitments and counterproductive government decisions. For example, the G20 has not yet fulfilled its 2009 commitment to end fossil energy subsidies. In 2021, global support for oil, gas, and coal still amounted to $11 million every minute, according to the International Monetary Fund. If G20 nations follow through, market forces will begin working to discourage investments in fossil fuels. A study in 2017 found that half of the fossil fuel investments in the United States would not have been profitable without tax breaks and subsidies.

Government relief during the COVID-19 pandemic was an example of significant public resources used in ways that undermine international efforts to cut carbon emissions. By mid-July 2020, the world's largest economies had committed at least $151 billion in bailout and stimulus money to fossil fuels with no environmental conditions attached. They allocated only $89 billion to clean energy. Under President Donald Trump, US bailouts awarded more than $120 billion in direct and indirect benefits to oil and gas companies, according to BailoutWatch, a nonpartisan program formed in 2020 to monitor recovery money.[21]

Mediation may become necessary if nations disagree about the application of new technologies. Solar radiation management (SRM) is an example. It would deploy mirrors in space or scatter aerosols in the atmosphere to reflect sunlight away from the Earth's surface. We don't know what effect this might have on photosynthesis, whether countries could weaponize it, or who would decide how and when to use it.

"SRM has world-altering potential," the Wilson Center points out. "Its governance needs to be early, anticipatory, and flexibly attentive to the range of possibilities that lie just over the horizon. . . . SRM demands consideration now." So does the tension between national sovereignty and global responsibility.[22]

DEMOCRACY'S BIG TEST

Speaking in the United States recently, UN High Commissioner for Human Rights Michelle Bachelet delivered a grim report on the state of democracy around the world:

> In 2021, the level of democracy enjoyed globally by the average person was down to 1989 levels. This means that democratic gains of the last 30 years have been greatly reduced. Last year, almost a third of the global population lived under authoritarian rule. And the number of countries leaning to authoritarianism is three times that of those leaning toward democracy.[23]

"It is a well-established empirical finding that democracies have declined in number and quality in recent years" while autocracies are evolving, Princeton professor Jan-Werner Muller writes in *Foreign Policy* magazine. The "flawed" rating for the United States indicates it could be one of them.[24]

The United States found itself at the precipice of authoritarianism during the Trump presidency, particularly on January 6, 2021, when the world watched thousands of his supporters riot in the US Capitol, hoping to help him retake the presidency despite losing the 2020 election. MAGA, the acronym for Trump's Make America Great Again slogan, fit nicely on baseball caps, but Trump did not define what he meant by "great." It turned out to be fascism. The Trump experience demonstrated how shockingly easy it is for fascism to take over even in the world's foremost model of democracy.

The rest of the world is watching to see whether the US public can figure this out. Reporting from a recent international youth conference in Germany, *Washington Post* columnist Christine Emba wrote, "In the eyes of much of the world, the United States' light has dimmed. . . . If we want to preserve our stature, we should begin to act. . . . The world is taking our decline seriously. It's time we did the same."[25]

However, preserving US stature isn't the country's most important objective. Its mission is, and always has been, to prove its second president, John Adams, wrong in his observation that "democracy never lasts long. It soon wastes, exhausts, and murders itself. There never was a democracy yet that did not commit suicide." In other words, America's mission is to prove that freedom is sustainable, even when severely tested, as it will be in the hotter times ahead.

I've described several problems with the US government, but the roots of gridlock are in society. In the past, US citizens have rallied against common enemies. Today, the enemy is their addiction to fossil fuels, but the nation hasn't yet rallied to the cure. In addition to the factors I've mentioned, narrow tribalism, manufactured outrage, the absence of a unifying national vision, and the loss of fundamental values are causing gridlock. While it's true we can't heal the climate without healing democracy, the United States can't heal its democracy until it fixes its social climate. How might it start?

Broaden Our Concept of Tribes

In a treatise on tribalism, psychology professors Dominic Packer and Jay Van Bavel wrote, "The underlying psychology of 'us and them' appears grounded in deep-rooted human tendencies to carve the world into groups and discriminate in favor of one's own." However, they point out that tribalism need not lead to conflict. Instead, it can be "the source of a much wider repertoire of actions, including cooperation, altruism, embracing diversity, and helping people radically different from ourselves."[26]

The inescapable reality is that we all belong to many tribes, each larger and more inclusive than the one before: friends, schools, neighborhoods, regions, nations, races, professions, political parties, and religions. We also belong to the human tribe and the tribe of sentient life on the planet. Curious about even larger tribes, we are searching for them in other solar systems and galaxies.

Several of our republic's weaknesses result from long-standing dissonances between tribes that think themselves superior to others rather than part of a larger unity. Racial prejudices are an obvious example, but the same dynamic prevents harmony between man and nature. Right now, learning to thrive and live harmoniously in the biosphere is considerably more important than starting over on Mars.

Restore Lost Values

A second threat to the "American experiment" is that fundamental values in its social contract have gone missing. We will not regain our coherence as a society, not to mention our civility, until we revive them.

Honesty means something. Facts mean something. Trust in each other means something. Character means something. Commitments mean something. Equality and fairness mean something. Freedom means something. Hard work means something. Courage means something. Harmony means something. Stewardship means something.

Many laws and other government interventions we chafe against are meant to compel compliance with behaviors that are second nature in a healthy society. The nineteenth-century philosopher and statesman Joseph de Maistre believed a wise and moral society did not need many laws. He pointed to the ancient Greek city-state of Sparta, which "justly boasted of having written its laws only in the hearts of its children."

Democracy means something, too. That's why more than one million US citizens have died for it since 1776. Unfortunately, it will soon be tested again, even harder, by epic problems we brought on ourselves with our conceited anthropocentrism and inattention to nature's limits. As the old truism warns, we get the leaders we deserve. They create the government we deserve, which shapes the policies we deserve and delivers the future we deserve.

We are capable of much better. We will achieve it if we reawaken the spirit President John Kennedy invoked about the challenge of putting a man on the moon:

> We choose to [do this] and do the other things, not because they are easy, but because they are hard. Because that goal will serve to organize and measure the best of our energies and skills, because that challenge is one that we're willing to accept, one we are unwilling to postpone, and one which we intend to win.[27]

Western civilization's idea of progress is culminating in a future filled with life-threatening and destructive surprises. Democracies must be much more robust and nimbler if they want to keep rights and freedoms in hotter times. Flawed democracies won't cut it. That's why we'd better identify the gridlocks and fix them now.

9 DEMOCRATIC GOVERNANCE FOR THE LONG EMERGENCY

Ann Florini, Gordon LaForge, and Anne-Marie Slaughter

If there was a moment when things might have gone differently for the planet, it was the late 1980s and early 1990s. In the preceding decade, the US Congress had passed legislation to protect endangered species, watersheds, and air quality. Scientists respected by the public were sounding the alarm about global warming, and there was bipartisan support for solving the problem—viewed as the sort of moon-landing-esque technical challenge that America excelled at. Before signing into law a cap-and-trade program to reduce acid-rain-causing sulfur dioxide emissions, President George H. W. Bush, a Republican, said, "A sound ecology and a strong economy are not mutually exclusive. They go hand in hand."[1] Three years later, President Bill Clinton and Vice President Al Gore proposed a federal budget with an ambitious environmental agenda that included a carbon tax.

Sensing an existential threat to its business, the fossil fuel industry mobilized. Its lobbyists persuaded senators to scuttle the carbon tax. Over the next several years, industry-funded PR and astroturfing campaigns sowed disinformation that undermined the scientific consensus and led the public to doubt the reality of climate change and to see environmentalism and economic growth as a zero-sum prospect. Climate change became tied to political identity in a new era of win-at-all-costs partisan bloodsport. Globally, the United States went from environmental leader to reactionary, slowing international action on climate change.

With the prospects of keeping warming below the 1.5°C target dwindling, we now face the certainty of a "long emergency," in which worsening wildfires, floods, extreme heat, and ecosystem destruction threaten to make much of the Earth unfit for human habitation. It will be a long and complex emergency. The environmental crisis is, like the biosphere itself, a complex system. Biodiversity loss, greenhouse gas emissions, soil depletion, coastal erosion—countless agents and processes all interact and adapt to feedback, rendering outcomes that are both greater than the sum of their parts and inherently unpredictable.

Moreover, the impacts of these outcomes vary greatly by geography, demography, and economy. In the United States alone, Louisiana faces flooding and hurricanes; California drought and wildfires. Farmers in Iowa may experience dwindling crop yields, while the imperative to shift to a non-carbon-burning economy will leave drill hands in Wyoming out of work. The impacts are similarly variable on a global scale; Canada and Russia can hope to obtain vast tracts of agricultural land, while many islands and low-lying countries are likely to disappear.

The combination of long time horizons and complex problems make for fractured and often paralyzed politics, as the US example reflects. Polling shows most Americans now support greater federal action to address climate change.[2] And yet—as is the case with many other issues ranging from gun control to immigration—the outsized power of special interests, minority rule, a toxic information landscape, a Supreme Court seemingly bent on eviscerating the government's regulatory capacity, and intractable structural features, such as first-past-the-post elections that deny representation in the House of Representatives to millions of Americans who live in districts that are firmly controlled by one party or the other, prevent the federal government from delivering on the will of the people.

To many American and global observers, the failure of the US government to address the environmental crisis is emblematic of the erosion of US democracy and its ability to solve problems. The conundrum of climate policy, however, goes far deeper. Even with revitalized institutions, the US federal government, an administrative state built on the foundation of an eighteenth-century constitution, would struggle to address the complex, interlocking challenges of climate change, biodiversity loss, and environmental degradation. It is a mechanical Newtonian system of checks and balances, action and reaction, force and counterforce that is too slow, static, selective, and closed to provide the energy, innovation, and collaboration that problem-solving in complex systems requires.

American democracy faces particular challenges, but most other democracies are similarly unprepared and unable to meet the needs of the planet in ways that will sustain their people. In a time of extraordinary social and environmental challenges, and facing the constraints of massive debt overhangs exacerbated by rapidly rising interest rates, all democracies need to rethink how they govern. They must restructure their systems to allow and encourage a move from government to *governance*, a set of interlocking institutions, networks, groups, and processes that is more open, participatory, inclusive, transparent, dynamic, and very, very messy.

From a governance perspective, it is possible to see the paralysis and inadequacy of the US federal government as a possible cause for hope rather than a counsel of despair. The federal *government* is not the only source of power, authority, and effective action. State and local governments, corporations, unions, nonprofits, media organizations, universities, faith groups, and the countless other associations that constitute the open US society all create policies and shape norms that affect humanity's impact on the biosphere. Federal regulation and legislation are vitally important, but they are only part of the solution—and a smaller part than our national discourse leads us to believe.

If humanity is to withstand the environmental crisis, then we need a bottom-up, all-hands-on-deck approach that empowers and engages all these diverse actors in the ongoing, long-term work of climate mitigation, adaptation, and drawdown. Governance systems can provide the flexibility and innovation we need, but they require shaping and pruning. As we will discuss in this essay, they must include structures, mechanisms, and norms that do three things:

- greatly increase participation and inclusion,
- facilitate the gathering and dissemination of all the information needed for that participation to foster effective governance,
- and communicate a positive shared vision.

We discuss each in turn. In the end, democracies include far more of the ingredients for successful governance than do other forms of government. But they will have to become far more democratic, with a small "d," and far more attuned to the demands of an age of complexity. Only thus can humanity achieve a sustainable existence within the bounds of the biosphere.

GOVERNANCE BEYOND GOVERNMENT

Governance is bigger than *government*. Governance structures can include governments, but they are typically wider, looser, and less formal.

Governments are the primary formal mechanisms for making and implementing decisions in a given place, and they have a special role: they can legally coerce people into doing what they say. Governance refers to any of the means by which a group of people set and enforce rules for themselves, or others. It is the infrastructure of collective action—the institutions, mechanisms, and information systems by which goals are set, information gathered and disseminated, and decisions made and implemented by and for groups of people.

Unlike democracy, governance is a neutral term.[3] A governance system need not have the consent of all, or even most, of the people affected by its rules and mechanisms. Even tyrants need ways of gathering information and implementing their whims. Moreover, governance need not involve governments at all. Corporations, nonprofits, religious organizations, universities, bowling leagues, book clubs, and families all have rule systems and shared understandings that translate individual preferences into group outcomes.

The forms vary widely. Large organizations might adopt governance that is centralized and top-down; dispersed across levels (such as in a federal system) or actors (in multi-stakeholder initiatives); or bottom-up (involving referenda, citizens' councils, or networks). Governance can also be informal, enforced through norms of behavior.

As technological change has accelerated and societies have grown more complex and interdependent, governance that serves the public interest has struggled to keep pace. In the United States, decades of underinvestment in and denigration of government, excessive freedom for corporations to pursue private interests, and social, economic, and technological changes that have hollowed out the middle class and weakened traditional associations have sapped governance structures of their ability to solve public problems.

But when it comes to the environmental crisis, the challenge is even more fundamental than that.

DEMOCRATIC GOVERNANCE AND THE LONG EMERGENCY

The environmental crisis is a complex, long-term crisis of a type humanity has never yet had to confront. It requires sustained attention over years and decades. It is predicted and yet unpredictable. It is a climate crisis, a biodiversity crisis, a pollution and waste crisis, all intersecting with one another. It requires holistic analysis and response from many different disciplines and perspectives.

By contrast, our existing governance processes are designed to handle problems that are relatively simple, isolated, and linear. Bureaucracies, for instance, sort all people and problems into boxes to which rote procedures can then be applied, in what Yale University scholar James Scott refers to as "seeing like a state."[4] In the name of efficiency, they sandpaper away the diversity that characterizes reality—an effective approach if the goal is to produce washing machines or welfare checks or soldiers, but an inadequate one for addressing a problem as complex, sprawling, and unpredictable as the environmental crisis.

These structures are ill-suited to the needs of the long emergency. They assume predictability, largely ignore the physical realities of the planet, focus on discrete problems without reference to the larger systems in which they are embedded, and assume equilibrium can be restored even after big shocks. They will fail to manage what is coming.

As the cascading crises of the long emergency accelerate, the temptation to turn to autocrats and populists will increase. Undemocratic governance may seem more efficient at managing crises than unruly democracies. Perhaps it can be, but if and only if the decision-makers have adequate knowledge of the details of all relevant issues, the means to ensure decisions are implemented as intended, and true concern for the public welfare over private interest—standards many autocrats are unlikely to meet.

Even if we were lucky enough to find a few benevolent and competent dictators, centrally directed crisis management wouldn't be enough. Undemocratic systems lack crucial mechanisms of resilience, which will be key for navigating the long emergency. Given that, in a complex system, conditions do not unfold predictably but rather emerge from the interactions of many different variables, timely feedback about local developments is critical for effective policy-making. In a top-down system, feedback from the powerless is invariably ignored until it's too late.[5] Only democracies tend to allow the space within which nonstate actors can engage, experiment, and self-organize.

That capacity for self-organization is critical because, in a world where the familiar patterns of nature are disappearing, we will need far more knowledge, far more capacity to act collectively—especially at local levels—and far better mechanisms for obtaining feedback on what interventions are working under what conditions. We will also need a shared vision of what we are trying to achieve—all with due attention to justice and fairness, given that virtually everyone will have grounds to feel aggrieved.

To that end, democratic governance amid the environmental crisis is not about finding the one "optimal" policy or a silver-bullet "best practice." Rather, it is about creating an enabling environment for widespread participation in inclusive governance, and for constant experimentation and learning.

PARTICIPATION AND INCLUSION

If, as argued above, existing bureaucratic processes are inherently insufficient for the task of implementing the degree of self-organization and experimentation that managing the long emergency requires, we need to change those processes. A first step is to make our governance bodies and processes far more participatory and inclusive.

One, we must develop new institutions that can harness and encourage climate action from as wide a range of participants as possible. The Paris Agreement took an important stride by bringing subnational governments, cities, companies, and other actors into the heart of the international climate action regime, one that previously included only nation-states. They are described in the agreement as "non-party stakeholders": they have no formal status in international law and thus cannot be parties to an international legal agreement, but they *do* have a direct stake in the preservation of the Earth's climate, *alongside* their governments. They can now participate directly in the international emissions governance regime, rather than only through the governments that purport to represent them. Their inclusion makes that regime more complex, but also potentially more effective. These actors can help catalyze more action, experimentation, and innovation at greater scale for climate mitigation.

But much larger steps are both necessary and possible. We need not just broader inclusion in existing types of organizations, such as treaty systems, but new institutional approaches. New markets that monetize ecosystem services and protect nature, for example, could direct the efforts and capital of millions of investors toward preserving and rehabilitating the biosphere—so long as they meaningfully involve public purpose advocates and the communities that live alongside those "investable" ecosystems.[6] New thinking about the purposes of for-profit business is creating space for business models that better align with the social and environmental necessities of the long emergency. New cross-cutting governance structures, already the focus of much experimentation, could expand to coordinate all the actors working on a particular environmental challenge.[7] Already, parts of the UN are experimenting with this approach: a food systems coordination hub anchored in the Food and Agriculture Office of the UN was launched last year.[8]

Two, we need to open up our existing governance bodies and processes to new participants and stakeholders—especially citizens and local communities. Many Indigenous and local communities depend on the environment for their livelihoods, and they will suffer the worst effects of the crisis. They have the most at stake and are the best positioned not only to provide input and oversight of environmental policies and measures, but also to organize and implement solutions. Nobel laureate Elinor Ostrom and many others have documented innumerable cases of local communities governing common pool resources sustainably and resiliently. Contrary to the orthodoxy that centralized, external control is the only way to avert the tragedy of the commons, collective, self-organizing, local-level governance arrangements have been effective for irrigation and rice farming in Bali, watershed management in the Andes Mountains, lobster fishing in Maine, and for preserving countless other common pool resources.

Governance bodies must give local communities meaningful voice when it comes to policy design, decision-making, grievance redress, and monitoring. Citizen assemblies, such as the one in France on climate change, can help ensure that governance reflects popular preferences and wins the trust of the governed.[9] Platform technologies such as those developed by the Cambridge-based company True Footprint, whose smartphone apps enable local volunteers to monitor development projects in challenging contexts, can reduce the transaction costs of involving large numbers of people in governance.

Many believe that a command-and-control state with legions of technocrats can more effectively enact policies than a messy, inclusive democracy. Proponents of this view point to China, which for decades has baked climate change into national planning and, unencumbered by gridlock and the whims of voters, has passed environmental policies mandating, among other things, renewable energy adoption and emissions reductions.

It's an open secret, however, that, especially in the absence of independent journalists and civil society watchdogs, local Chinese officials underreport coal consumption and businesses falsify emissions data.[10] The Chinese government is cracking down, but heavy-handed central planning in a fluid, complex system can generate unintended risks and consequences. Mao Zedong's 1958 call for citizens to kill all sparrows—deemed crop-destroying pests—succeeded in wiping out an estimated two billion birds.[11] But the ensuing insect infestations destroyed agriculture and contributed to the Great Famine, in which tens of millions of people died. More recent decrees—such as the one-child policy, which curbed China's population but now puts it in an economic-demographic bind of growing old before it grows rich—also illustrate the inflexibility of autocracy to quickly course-correct, an imperative when dealing with complexity.

Involving new stakeholders in governance bodies can help trigger action and disrupt a harmful status quo. For instance, steps to open up corporate governance—such as by removing dual-class shares that entrench the decision-making power of founders on corporate boards—can enable new voices to push changes in line with majority preferences. The election of activist shareholders committed to climate action onto the boards of oil majors, such as the three who attained seats on the ExxonMobil board in 2021, is a promising illustration of this approach. Another positive development is the rapid growth of public benefit corporations, businesses that bake in a fiduciary duty not only to their shareholders but also to the good of the general public, including their employees and their community.[12]

INFORMATION AND TRANSPARENCY

Expanded direct participation in governance will not work without expanded information. To foster the capacity to manage climate-linked disasters, cope with cascading shocks, end environment-harming practices, and plan for a volatile future, participants will need to be fully informed. That means we need not only to make information open, accessible, and digestible, but also to increase nongovernment participation in designing and populating the information systems that inform governance. Otherwise, we lack insight into what data to collect, much less how to use information to address our problems.

Most of the conversation around the role of information in democratic systems focuses on transparency: compelling governments and corporations to produce and disclose information about activities that affect the public in ways and formats that are accessible to as many people as possible. Transparency is essential for good governance. It is key to accountability, and accountability is the means by which representation via elections is rendered truly democratic.

But the role of information in democratic governance is much bigger, and raises a host of much deeper questions:

- What data are collected in the first place, and why?
- To whom are the data disseminated, and in what format?
- Whose knowledge, beliefs, and judgment are applied to the data in figuring out what the data mean?
- Under what conditions does information foster effective collective action?

All of these require attention. For example, it is clear that, despite the growth of substantial new research aimed at providing data about the environment and about the impacts of volatility, governmental capacity to provide all of what is needed remains inadequate. Increasingly, however, governments are not the only source of such data, with a huge array of nongovernmental actors looking to fill the gaps. One environmentally minded NGO, the Environmental Defense Fund, is even building its own observation satellite (MethaneSat) to monitor methane emissions, an enormous gap in the existing climate monitoring infrastructure. Such efforts are crucial, but insufficient.

Not only do we need information systems that combine huge quantities of Earth systems data and human systems data (much of which we do not yet even collect), but if any of that is to inform governance and spur collective action, then we need to translate it into terms that are meaningful for policy-making and public discourse. A major success of the IPCC was to demonstrate the scientific rationale for politically agreed degrees of warming—1.5°C, in particular—as the target around which to organize international commitments. Other consequential aspects of the environmental crisis—biodiversity loss, desertification, habitat destruction, to name a few—lack such targets.

We also need to rethink our economic indicators and targets, which right now reflect, and encourage, policy-making obsessed with growth. National GDP figures and stock prices are poor measures of sustainability, not of much use in determining what is salvageable of a coastal city or finding ways to protect the productive capacity of a local farming community battered by incessantly volatile weather.

Most information systems are mediated through institutional intermediaries, primarily government agencies. Usually, agencies use data to support their own regulatory work, but provision of information to the interested public is itself a key governance function. Sometimes providing information is all a government needs to do: rules that require restaurants to post their hygiene ratings or airlines to post the on-time arrival records of specific flights give consumers what they need to make informed choices—and thus influence business behavior using the wonderfully efficient signals of market demand.

But market prices do not incorporate most climate-relevant information, since carbon pollution is rarely priced into goods and services. Indeed, most sustainability metrics are a mess, either far too complex to be actionable or too simplistic to provide good feedback—when they are not actually designed to mislead investors or consumers.

And information can foster effective collective action only when the source is trusted. With Americans' trust in government and traditional media outlets historically low, rebuilding trust in information sources is a major challenge for effective governance.[13] Here, too, inclusion and participation are key ingredients. If you or someone close to you whom you trust are part of the system and can see where information is coming from, you're more likely to trust the information.

But as has been all too evident in recent years, it matters greatly whether the trusted sources are in fact trustworthy. Conspiracy theories and fake social media news thrive precisely because they feel more participatory, personal, and inclusive. Information systems have to be designed to ensure that the basic data are accurate, the information extracted from those data are of value, and the information is interpreted using beliefs and judgment systems that are rooted in reality.

SHARED STORIES

Generating the society-wide collective action needed to address the environmental crisis will require more than just improved governance structures and systems, even with all the inclusion and transparency we can muster. Perhaps above all else, it will take shared stories. The historian Yuval Noah Harari writes that "any large-scale human cooperation . . . is rooted in common myths that exist only in people's collective imagination."[14] Human progress is the story of groups of people imagining new realities and then enrolling enough people in those visions to bring them into being.

A shared vision is especially important for a democratic state, which lacks autocracy's power to force people to act. A compelling story can motivate free individuals to abandon a status quo and embrace new norms, behaviors, and goals. It was not the technical feasibility of rocketry or the security calculus of the Cold War that rallied the American public and government to devote time and resources to the moon mission; it was the aspirational narrative of exploration and exceptionalism voiced by John F. Kennedy. Democracy itself was born of an inspiring, empowering idea—that the people, not a divinely ordained monarch, had an inalienable right to self-government.

Today, US democracy struggles to articulate collective stories. Partisan polarization, cable TV, the Internet, and social media have fractured the information environment. We have opposing narratives not just about who we are and what we want to accomplish, but about what exists: the facts of reality itself. In this landscape, even attacks and national emergencies—Russia's interference in the election; the COVID-19 pandemic; the January 6 insurrection—fail to muster a collective response. America contains multitudes, but rarely in the nation's history have the narrative threads that weave us together been so frayed.

Environmental action is a casualty of this phenomenon. For years, the fossil fuel industry and the Republican Party intentionally undermined the scientific consensus, thwarting a shared understanding of the crisis. But even with widespread agreement that the planet is warming, species are dying, and action is needed, we lack a common vision for a sustainable, prosperous future.

Creating that vision will require dispelling the two narratives that currently dominate the discourse. The first is that saving the environment means harming economic growth. George H. W. Bush's dictum that "a sound ecology and a strong economy . . . go hand in hand" gave way to the belief that climate action would kill jobs and raise consumer prices.

Yet research has shown that emissions reductions and economic growth are not mutually exclusive; from 2000 to 2014, thirty-five countries cut CO_2 and increased their GDP.[15] As the success of valuable companies such as Tesla demonstrates, the transition to renewable energy will create jobs and industries that will benefit the economy—especially when considered against the costs of inaction. A January 2022 Deloitte report found that insufficient action on climate change would cost the US economy $14.5 trillion over the next fifty years, whereas rapid decarbonization would add $3 trillion and almost one million more jobs than the status quo.[16] Swiss Re, one of the world's largest insurers of insurance companies, reported in 2021 that unmitigated climate change would reduce global economic output 11 to 14 percent by 2050, costing the global economy the equivalent of $23 trillion.[17]

The second prevailing narrative is that we are doomed. The environmental crisis is such a massive problem, the opposition of special interests so powerful, and the inertia of average citizens so great that even our best efforts will fall short. Catastrophe approaches, and we are powerless to stop it. Though a clear-eyed view of the reality of the challenge is critical, cynicism and resignation are unhelpful. Fear and despair will slow action and alienate potential supporters.

We need an inspiring, broadly inclusive, positive-sum vision of a sustainable future. We need to speak about the environment in a way that invites participation, showcases opportunities, and gets past entrenched biases. For instance, "climate change" is a freighted term tied to partisan identity. But "nature" is more neutral and attractive.[18] A sportsman in Idaho might doubt the science of global warming but totally support preserving the natural wilderness where he hunts and fishes.

Rather than centering the discourse around national politics, we need to amplify solutions from across sectors and segments of society. These might include new sustainable agriculture practices, nature-based markets, endangered species reintroduction, or carbon capture technologies. Doing so can have powerful demonstration and inclusion effects that encourage imitation and adoption.

Part of the power of stories is their role in shifting norms of human behavior. In 1965, 43 percent of American adults smoked cigarettes, a behavior that was perceived as cool and mainstream. In 2018, only 13 percent of adults smoked, after public awareness campaigns highlighting the health effects of smoking prompted a shift in the social narrative. It might seem hard to imagine, but similar shifts could reduce a variety of harmful behaviors—car ownership, eating meat, single-use plastic consumption—that right now seem like immutable features of modern life.

These principles of participation, transparency, and storytelling provide key criteria for evaluating and (re)designing both existing and possible new institutions of governance: Who participates and how? What information/transparency/feedback mechanisms are included? What stories are the focal points? But these principles also offer much more.

If our governance systems can meaningfully increase participation, foster transparency, and promote a shared positive vision, those changes will not just enable us to address the environmental crisis. They will also strengthen democracy itself. These measures will help restore trust and connectivity across society. Two hundred years ago, Alexis de Tocqueville recognized that social capital—the capacity to form voluntary associations—was the essential ingredient of US democracy. He was focused on autonomous individuals coming together for a common purpose, underscoring that, equally important for a society that has often tilted too far in the direction of individualism, rugged or otherwise, are the connections to other human beings—family, friends, teachers, advisers, mentors, caregivers, and others—that allow individuals to grow and flourish. Those connections also promote a larger sense of interdependence and solidarity.

Social capital is also the foundation of resilience. Researchers have found that the single best predictor of how well and how quickly a community withstands and bounces back from a natural disaster is not per capita GDP or emergency response infrastructure or even advanced planning, but rather social capital, the strength and density of relationships in a community. When disaster strikes, people turn to their neighbors for help and support.

That social capital will be crucial for adapting to the impacts of the environmental crisis is a cause for concern. For, over the past few decades, researchers and journalists have documented the decline of social capital in American society.[19] From the disappearance of bowling leagues to dwindling church attendance to the rise in social media and online communities at the expense of in-person interaction, the practice of voluntary association in US life is shifting and diminishing. We have more opportunities to connect and engage than ever before, and yet, Americans—especially youth—report feeling increasingly isolated, lonely, and disconnected.[20]

Responding to the environmental crisis is the challenge of a generation. But for all its negative effects, it can also—will also—present an opportunity for renewal. Our governance systems can help engage all segments of society in the fight—and in so doing, help restore the trust and connectivity that constitute the fabric of American democracy.

10 CAN THE CONSTITUTION SAVE THE PLANET?

Katrina Kuh and James R. May

Fossil fuels were the toast of the town. For its centennial in 1959, the American Petroleum Institute threw itself a party at Columbia University in New York City called "Energy and Man." All the Big Oil bigwigs were there, representing most worldwide fossil fuel production. The celebration culminated with a not-to-be-missed keynote speech by Edward Teller, a political conservative largely credited with being the "father of the hydrogen bomb." Teller was a celebrated scientist, and *Time* magazine named him "Man of the Year" in 1960. It was a standing-room-only event.

But to everyone's astonishment, Teller wasn't in a cheering mood. After clearing his throat, he said,

> Ladies and gentlemen, I am to talk to you about energy in the future. This, strangely, is the question of contaminating the atmosphere. Whenever you burn conventional fuel, you create carbon dioxide. The carbon dioxide is invisible, it is transparent, you can't smell it, it is not dangerous to health, so why should one worry about it?

Teller was worried in 1959 when human-caused atmospheric CO_2 loadings were about nine billion metric tons per year, less than one-quarter of what they now are. Moreover, since 1959, the concentration of CO_2 in the atmosphere has increased from 315 to more than 420 parts per million, and the Earth's average temperature has risen to 57°F from 51°F. As explained elsewhere in this volume, the causes and effects are unprecedented.

Teller continued:

> Carbon dioxide has a strange property. It transmits visible light but it absorbs the infrared radiation which is emitted from the earth. Its presence in the atmosphere causes a greenhouse effect. It has been calculated that a temperature rise corresponding to a 10 per cent increase in carbon dioxide will be sufficient to melt the icecap and submerge New York. All the coastal cities would be covered, and since a considerable percentage of the human race lives in coastal regions, I think that this chemical contamination is more serious than most people tend to believe.[1]

The underlying processes were not hard to grasp; in 1958, a popular children's science program directed by Frank Capra, who studied chemical engineering at Caltech before becoming a movie director, explained the relevant physical and chemical mechanisms.[2] And Teller's "more serious chemical contamination" was no surprise to the federal government.[3] Within a decade, policymakers like Senator Daniel Patrick Moynihan had sounded alarms. The summer of 1988 was dominated by congressional hearings about climate change. Big Oil knew, too. After a five-year investigation, the Philippines Human Rights Commission recently determined that that Big Oil suppressed climate science for half a century, and then lied about it, repeatedly.[4] Avoidable apocalypse awaits, even the Ninth Circuit agrees.[5]

Yet here we are: No national, economy-wide emission limits. Increasing emissions (again). Not a single case that's yet reached the merits of the causes of climate change, who knew what and when, who is responsible, and what to do. No one has gone to jail. No executive has been fined. No oil lawyer has been disciplined.

Even a healthy US democracy would struggle mightily to respond effectively to the "wicked" problem of climate change. As humans, we would have to overcome innate cognitive limitations to accept the connection between everyday actions like driving and attenuated effects on climate. As voters, we would have to appreciate the importance of addressing a problem with limited immediate salience primarily for the benefit of future generations. Politicians would need the courage to fight for policies that would yield no visible benefits for voters in their political lifetimes. Change would have to occur over the tooth-and-nail opposition of the fossil fuel companies, among the most powerful corporate special interests in the history of our country.[6] And we would need to be open to transformative change and a rethinking of values and priorities to develop a shared vision for a future radically different from settled expectations, all while reckoning with the ways that historical injustice—most notably colonialism and racism—produced and remain embedded in the systems that brought us to this precipice.

And our democracy is decidedly not healthy. Corporations pour money into elections virtually unrestrained and face no consequences for flagrantly lying to the public—most notably for present purposes about whether climate change is real and dangerous and what causes it—to maximize their profits. The right to vote is now more theoretical than real, with federal protections dismantled and many state and local governments adopting measures that make it harder (especially for disfavored people) to vote. The votes of those who make it into the voting booth are counted—our electoral system is sound. But it is not perceived as such; ideological fantasies of voting fraud cause wide swaths of the public to reject legitimate electoral outcomes. And myriad pathologies in the information environment, from social media algorithms to purposeful disinformation efforts, have fractured our ability to talk to and reason with one another.

Nor is our climate system healthy or stable. We already limp from unprecedented crisis to unprecedented crisis—the Pacific Northwest heatwave, deadly fires, a hurricane that makes landfall in Louisiana and drowns eleven people in New York City basement apartments—with hardly a chance to catch our breath in between. This rapidly changing physical reality adds further stress to our ailing democracy. We face the daunting task of attempting to transform our society, law, and economy to cease emissions and sequester carbon while also transforming our society, law, and economy to adapt to ongoing and worsening dislocations from climate change.[7] In short, our democracy must somehow martial focus, purpose, and resources to address climate change's long-term challenges to life and justice while enduring increasingly frequent and extreme climate-related events that take lives, exacerbate injustice, and put unprecedented strain on democracy and governance.

The first two sections detail how and why the Constitution is hostile to climate policy. The third section explains why courts are reluctant to engage climate change. The fourth section explores what it means for our brand of democracy.

HOW THE CONSTITUTION OBSTRUCTS CLIMATE POLICY

How and why does the constitutional status quo fail to meet the climate crisis? It begins with constitutional design. The Constitution grants the SCOTUS, and such lower courts as Congress establishes, "judicial authority." The SCOTUS was thought to be the "least dangerous branch," and a less prestigious appointment than to, say, supreme courts in Virginia or New York, among the reasons some of the first appointed Supreme Court justices, Jay and Rutledge, quit the bench. But the import of judicial review was transformed in *Marbury v. Madison* when Chief Justice John Marshall famously declared that it is "emphatically the province and the duty of the judicial department to say what the law is," as it has in more than twenty thousand decisions since its inception in 1789.[8]

The US Supreme Court has nine members, which it did during the 1959–1960 term when Teller gave his remarks. Twenty-eight more people have been appointed since then. It has had 115 members in all. Yet very few have acknowledged that catastrophic climate change is both happening and caused by humans. And even those that acknowledge climate change don't believe there is anything courts can do about it.

The US Constitution's 7,369 words all but ignore environmental concerns. It was crafted to address separation of powers, federalism, and civil liberties. Here we provide a taxonomy of the sources of and limits to federal and state authority to reach climate change. Virtually nothing is uncontested, giving rich context to Chief Justice John Marshall's maxim that, "we must never forget that it is *a constitution* we are expounding."[9] But, forget it or not, the Constitution has failed the climate. To understand how and why, we explain that myriad constitutional provisions relate to climate change and, in many cases, constrain climate action. Taken together, these provisions contribute to an often unacknowledged yet critically important shared understanding of the power to make and the permissible content of climate policies within our constitutional framework. These are the constitutional shoals to navigate, for example, for an answer to Daniel Lindvall's query about whether democracy can safeguard the rights of future generations, as well as to achieve the climate-effective, bottom-up, all-hands-on-deck Earth-systems governance endorsed by Ann Florini, Gordon LaForge, and Anne-Marie Slaughter elsewhere in this volume.

Sources of and Limits to Federal Regulation of Climate Change

Sources

The US Constitution fails the climate seemingly at every turn. The Commerce Clause provides that "Congress shall have the power . . . to regulate Commerce . . . among the several states." The court has held that the Commerce Clause permits Congress to regulate in three areas: channels of interstate commerce (such as navigable waters), instrumentalities of interstate commerce, and activities that "substantially affect" interstate commerce.[10] The court then described the "substantially affect" test as a function of whether (1) the underlying activity is "inherently economic," (2) Congress has made specific findings as to effect, (3) the law contains a jurisdictional element, and (4) the overall effects of the activity are actually substantial.[11] Seldom in applying this vigorous analysis have courts found Congress lacks constitutional authority to protect fish, flies, spiders, "hapless toads," waters or wolves that exist solely within a single state. Moreover, the court has upheld congressional authority when Congress could have a "rational basis" for concluding that "activities, taken in the aggregate, substantially affect interstate commerce."[12] Yet, despite this wide authority, there are those who question whether Congress has authority to address climate change because climate change is not an inherently economic activity (selling oil is, but not changing the climate).

The Treaty Clause provides that the executive branch "shall have power, by and with the advice and consent of the Senate, to make Treaties, provided two thirds of the Senators present concur." After a treaty is approved, Congress has the power under the Necessary and Proper Clause "to make all laws which shall be necessary and proper for carrying into execution . . . all . . . powers vested by this Constitution in the Government of the United

States." Because of the Supremacy Clause, these laws effectively preempt any conflicting laws enacted by states, although states are inclined to argue that the Tenth Amendment's reservation of power "to the states . . . or to the people" limits the federal reach of statutes on matters traditionally left to the states.[13] While the Senate ratified the UN Global Climate Change Convention in 1992, Congress neglected to enact implementing legislation. While the US State Department under the Clinton administration signed the emission-reducing 1997 Kyoto Protocol, the US Senate tabled it. The Senate ignored the 2015 Paris Agreement altogether.

The Spending and General Welfare Clauses permit Congress to tax and spend to "provide for the common defense and General Welfare of the United States," by attaching conditions to the receipt of federal funds, provided the conditions are not coercive,[14] authority that has been used to implement various federal environmental laws. Such "cooperative federalism" is the bulwark of most of the nation's health and welfare legislation, including environmental protection, but again, not about climate change.

The Property Clause authorizes Congress to make all "needful" rules concerning federal land, which constitutes about 28 percent of the country. The court has interpreted the scope of this authority to be "virtually without limitation."[15] With this authority, Congress has enacted numerous laws allowing for development, use, and exploitation of natural resources on federal lands. Other provisions of the Constitution have allowed Congress to exercise relatively unquestioned authority to protect natural resources on federal enclaves under the Enclave Clause.[16] Congress also has authority to "acquiesce" to presidential power to "reserve" natural resources on federal land.[17] Yet, again, these constitutional authorities are generally used to develop and not diminish fossil fuel production, as the Biden administration's responses to the Russian oil embargo demonstrate.

Limits

The Tenth Amendment provides that "the powers not delegated to the United States by the Constitution, nor prohibited by it to the States, are reserved to the States." Tenth Amendment jurisprudence has curtailed environmental programs that upset political accountability and diminish state dignity.

The court has held that Congress may not "commandeer" state political or personnel resources by requiring a state to "take title" of its own low-level radioactive waste even if it fails to arrange for proper disposal under federal law.[18] In the climate context, this jurisprudence has all but foreclosed using state resources to implement climate policies.

The Eleventh Amendment provides that "the Judicial power of the United States shall not be construed to extend to any suit in law or equity, commenced or prosecuted against one of the United States by Citizens of another State." Absent express consent, states are immune from accountability under federal law, in federal court,[19] in state court,[20] or before federal agencies.[21] This again ties Congress' hands in holding states to account for climate change. State officials are still subject to federal causes of action for prospective injunctive relief, a tack left available under *Ex parte Young*.[22] Still, no jurisprudence yet exists to hold state officials to account for climate change.

The Fifth Amendment forbids the government from "taking private property for public use without just compensation." "Taking" includes so-called regulatory takings that arise when regulation goes "too far."[23] This involves a balancing approach that turns on how closely the impact of the challenged regulation resembles a physical occupation of the regulated property. In so doing, courts weigh three factors to determine whether a government regulation triggers the obligation to compensate the property owners: (1) "the economic impact of the regulation on the claimant," (2) "the extent to which the regulation has interfered with distinct investment-backed expectations," and (3) the "character of the governmental action," that is, whether it amounts to a physical invasion or merely affects property interests through "some public program adjusting the benefits and burdens of economic life to promote the common good."[24] The court has held that depriving a landowner of all economically viable use of property, requiring the owner to maintain a public pathway to the beach,[25] or setting aside land for a greenway along a nearby creek[26] goes too far, unless such use constitutes a nuisance under the state's traditional common law.[27] Moreover, even preexisting state laws can constitute a compensable taking.[28] The prospects of extensive compensation make governments shy about progressive climate programs.

The Due Process Clause of the Fifth and Fourteenth Amendments prevents the government from "depriv[ing] any person of life, liberty, or property, without due process of law." Substantive due process secures "fundamental rights," which the court has never interpreted to include anything about the environment.[29] Almost without exception, however, courts have declined to find that individuals have a substantive right to a stable climate.[30] The procedural prong of the Due Process Clause also requires sufficient process associated with individualized decision-making associated with deprivation of a constitutionally protected interest, and that any punitive damages be proportional to compensatory damages, both of which tap the brakes on progressive climate regulation.

The Equal Protection Clause of the Fourteenth Amendment provides "nor shall any state deny to any person within its jurisdiction the equal protection of the laws." The Supreme Court has interpreted this to require evidence of "invidious" express or intentional racial discrimination to warrant heightened scrutiny to discriminatory governmental action.[31] Thus far, no court has held that climate policies contravene equal protection.

The Privileges and Immunities Clauses provide that "The Citizens of each State shall be entitled to all Privileges and Immunities of Citizens in the several States" and that "No State shall make or enforce any law which shall abridge the privileges of immunities of the citizens." While several current SCOTUS members have expressed an interest in reviving it,[32] the US Supreme Court has largely read the clause out of the constitution for modern applications, making it unavailable for addressing climate change.[33] Likewise, the court has never construed the Ninth Amendment's text that "the enumeration in the Constitution, of certain rights, shall not be construed to deny or disparage others retained by the people," as affording a right to a healthy environment or stable climate.

The Nondelegation Doctrine stems from Article I of the Constitution, which vests "all legislative" authority in Congress, and presumably not in agencies charged with implementing national policies. Nonetheless, while Congress may not "delegate" legislative authority to agencies that administer federal law, legislation that provides an "intelligible principle" to guide the exercise of agency discretion will be upheld. The court declined to use the doctrine to strike a provision of the CAA that charges EPA with the duty to set national ambient air quality standards as "requisite" to protect public health and welfare.[34] Yet echoes of the doctrine can be seen in rulings that are skeptical of whether the Clean Air Act's definition of "air pollutants" includes greenhouse gases.

The Political Question Doctrine holds that matters demonstrably committed to a coordinate branch of government, or that lack ascertainable standards, or that could otherwise result in judicial embarrassment are nonjusticiable.[35] The court has recognized that executive powers over foreign affairs, impeachment, and treaty abrogation are political questions into which courts "ought not . . . enter [the] political thicket."[36] The court has declined to engage arguments inviting analysis under the Political Question Doctrine in holding that the federal CAA provides EPA with authority to regulate emissions of GHGs from new motor vehicles.[37] Nonetheless, several federal courts have turned recently to the Political Question Doctrine in deciding that cases involving climate change are nonjusticiable.

The Displacement Doctrine is grounded in separation of powers. It stands for the proposition that federal law enacted by Congress or implemented by the executive branch can displace the role that federal courts have to hear (at least) federal common law causes of action. The court has held that the discretionary authority that the Clean Air Act provides to EPA to regulate greenhouse gases, coupled with the corresponding regulatory actions EPA has taken, displaces the federal common law for public nuisance actions brought under federal common law concerning climate change.[38]

The Standing Doctrine has had a pervasive and deeply imbedded influence on environmental law. Article III extends "judicial authority" to "Cases . . . and Controversies." In general, the Supreme Court has construed this provision to require that a plaintiff show a personal injury that can be traced to the defendant's conduct and redressed by a judicial remedy.[39] The court recognized noneconomic, aesthetic, and environmental interests as legally cognizable "injuries" that can serve as a sufficient basis for constitutional standing under Article III,[40] and made clear that it is injury to a person, and not the environment, that matters, thus obviating any need to show environmental degradation to support constitutional injury.[41] States are entitled to "special solicitude" in standing analysis in cases involving state efforts to protect natural resources.[42] There, the court recognized Massachusetts's potential shoreline loss as a legally cognizable injury in allowing it to challenge EPA's failure to regulate GHG emissions. Individuals, on the other hand, must still show a tight "geographic nexus" between the claimed injury and the federal action.[43] Thus, standing remains a tough obstacle to plaintiffs in climate cases.

Sources of and Limits to State Regulation of Climate Change

Sources

The Tenth Amendment "reserves" state authority in areas neither reserved for Congress nor withheld from the states. Many states have adopted extensive laws governing the environment, especially where federal regulation has left gaps. Yet states generally have avoided enacting laws that limit GHG emissions, largely owing to limitations on state authority discussed below.

The Compact Clause of the US Constitution provides that "no state shall, without the consent of Congress . . . enter into any agreement or compact with another state." Historically, water resource allocation has been the area where regional issues warranted an appreciation of the Compact Clause. The Regional Greenhouse Gas Initiative is a cooperative, market-based effort among eleven states to reduce GHG emissions. Yet Congress hasn't approved the compact, and it isn't enforceable.

Limits

The Dormant Commerce Clause is most often used to describe limits on a state's authority to adopt laws or policies that discriminate against interstate commerce because they favor one state or impose an excessive burden on outsiders. For example, the court has struck down a ban on the importation of dangerous out-of-state waste,[44] higher tipping fees or surcharges for wastes generated out of state,[45] and waste flow control ordinances prohibiting landfill operators from accepting out-of-state waste or requiring all county waste be processed at the county's facility,[46] and a state statute that restricted withdrawal of groundwater from any well in the state for use in an adjoining state,[47] a state law that prohibited the export of energy generated within the state,[48] and other state initiatives awarding tax credits for in-state ethanol production[49] and requiring that in-state power plants burn in-state-mined coal.[50] The body of case law here suggests obstacles for states aiming to reduce GHG emissions by promoting in-state renewables, for example.

Public facilities that regulate the environment for public benefit may enjoy wider latitude under the Dormant Commerce Clause; for example, a flow control ordinance that required that all solid waste generated within the county be delivered to the county's publicly owned solid waste processing facility does not violate the dormant Commerce Clause.[51] Yet energy companies are privately owned, even if pervasively regulated.

The Supremacy Clause provides that "the Constitution and the Laws of the United States which shall be made in Pursuance thereof . . . shall be the Supreme Law of the Land." Absent specific intent to preempt, the modern court has held that Congress may preempt state law implicitly by "field" preemption, when Congress occupies a field of interest so pervasively that preemption is assumed, or when state law "conflicts" with federal law. The court has held that a comprehensive regulatory scheme involving environmental protection may occupy the field and thus implicitly preempt state common law.[52] Thus, preemption limits the availability of private causes of action (e.g., public or private nuisance) to address climate change.

CONSTITUTIONAL OMISSION

The Constitution presumes the existence of an environment capable of supporting a flourishing society but does not explicitly recognize or protect the environment. Indeed, it was historically uncertain whether the Constitution empowered the federal government to act to protect the environment at all.[53] The textual silence of the Constitution with respect to the environment is mirrored by judicial silence about how the environment supports other rights explicitly recognized in the Constitution. Despite the obvious fact that there can be no life or liberty without functioning ecosystems, courts in the United States do not recognize any federal constitutional environmental rights, even to the extent an environmental right might be deemed appurtenant to explicitly enshrined constitutional rights. The constitutions of many countries explicitly protect the environment, and in many others, courts have interpreted nonenvironmental constitutional provisions to necessarily include environmental rights. That is not the case in the United States. The Constitution does not explicitly protect the environment nor is protection of the environment recognized as required to protect other constitutional rights.

The absence of clear and broad constitutional authority to protect the environment—explicit in the text of the constitution or understood by courts to reside within other enumerated powers—limits the scope of federal environmental law. One aspect of this limit is foundational. It required judicial willingness and a stretching of doctrine to find constitutional authority on which to adopt our core federal environmental laws. What might the laws coming out of our great public awakening to modern environmental problems in the 1970s—the National Environmental Policy Act, the Clean Air Act, the Clean Water Act, the Endangered Species Act—look like had they been adopted against a backdrop of clear and broad constitutional authority?

The dearth of constitutional authority limits not just the structure of our core federal environmental statutes, but also the way that they are interpreted and applied. Internalized understandings of the limits of what is constitutionally possible constrain our policy-making imagination when it is imperative that we effect systemic, transformational change to respond effectively to today's environmental challenges.

And the absence of clear and broad authority to protect the environment is merely one side of the coin. In addition to failing to give the federal government clear power to *protect* the environment, the Constitution also fails to impose clear limits on the federal government's power to *harm* the environment, let alone impose a duty on the government to prevent harm to the environment. The Constitution's failure to identify limits on government harm to the environment compounds the lack of express power to act to protect the environment. The Constitution is (at least as a matter of express text) silent when it comes to protecting individuals from government destruction of the ecological necessities for healthful life. And, unlike in the case of constitutional power to legislate on environmental matters, courts have been largely unwilling to read such limits into the Constitution.

To appreciate how constitutional silence on the environment contributes to constitutional hostility to environmental protection, it is important to recall that the Constitution affords other values—private property, speech, states' rights—explicit recognition and protection. The constitutional omission of the environment thus diminishes the constitutional importance of environmental interests when considered relative to other interests explicitly protected by the Constitution. One-sided constitutional protection for private property, for example, hobbles adaptation policy by dissuading governments from restricting new development in climate-vulnerable locations out of fear of triggering an obligation to compensate private landowners.[54]

The textual constitutional omission of the environment could be overcome by judicial interpretation. For example, during the Progressive Era, scientists, attorneys, and politicians succeeded in persuading courts to interpret the Commerce Clause to give the federal government significant constitutional environmental authority. And there are many powerful arguments that, despite the lack of explicit text and a historical doctrinal focus on the Commerce Clause, the Constitution can and should be understood to afford broader environmental powers to and impose environmental limits and duties on the government. To date, however, courts have largely abdicated their institutional role in the development of climate policy.

JUDICIAL ABDICATION

Advocates, alarmed at the closing window for mitigation to avoid catastrophic warming, have repeatedly beseeched the courts to use their constitutional authority to compel or prompt more meaningful mitigation policy. Courts decline, insisting that our constitutional structure (the separation of powers among the legislative, executive, and judicial branches of government) renders climate change policy exclusively a matter for the elected branches. In doing so, the courts hold up the need to respect democracy, positing that climate change policy is so complex and central to the polity that decisions about it should not and cannot be made by unelected judges. But this view abdicates the essential role of courts in our constitutional democracy to protect rights, correct for pathologies that subvert the political process, and engage in conversation with the other branches.

Courts have declined to even consider whether the federal government has a constitutional duty to its citizens to prevent (or at least not inflict) climate harms; has blocked those harmed by climate change from seeking relief directly from emitters and fossil fuel producers under the federal common law; and has rejected interpretations of the Clean Air Act that would allow for meaningful federal mitigation under existing law. Although the decisions technically reside within distinct legal doctrines (standing, displacement, statutory interpretation), substantively the decisions all reflect the judiciary's conviction that it would offend the constitutional structure for courts to engage on climate change policy because in so doing courts would overstep their constitutional role in our system of separated powers.

For example, a federal appellate court held that the *Juliana* plaintiffs—children whose lives and futures have been upended by climate change—did not have standing to challenge federal climate policies because

> It is beyond the power of an Article III court to order, design, supervise, or implement the plaintiffs' requested remedial plan. As the opinions of their experts make plain, any effective plan would necessarily require a host of complex policy decisions entrusted, for better or worse, to the wisdom and discretion of the executive and legislative branches.[55]

This is especially evident in the climate context. In holding that federal common-law-based claims are displaced by federal law, the court reasoned,

> It is altogether fitting that Congress designated an expert agency, here, EPA, as best suited to serve as primary regulator of greenhouse gas emissions. The expert agency is surely better equipped to do the job than individual district judges issuing ad hoc, case-by-case injunctions. Federal judges lack the scientific, economic, and technological resources an agency can utilize in coping with issues of this order.[56]

Yet, when asked to allow agencies to interpret the Clean Air Act to require significant reductions in emissions from those same power companies, the courts hesitate, fretting that robust regulation of the emissions from existing power plants would have significant societal impacts and therefore requires additional and clear congressional authorization:

> Capping carbon dioxide emissions at a level that will force a nationwide transition away from the use of coal to generate electricity may be a sensible "solution to the crisis of the day." . . . But it is not plausible that Congress gave EPA the authority to adopt on its own such a regulatory scheme in Section 111(d). A decision of such magnitude and consequence rests with Congress itself, or an agency acting pursuant to a clear delegation from that representative body.[57]

Additionally consequential is the current court's interest in curtailing unenumerated rights, such as to abortion: "The Constitution makes no reference to abortion, and no such right is implicitly protected by any constitutional provision [as] any such right must be 'deeply rooted in this Nation's history and tradition' and 'implicit in the concept of ordered liberty.' The right to abortion does not fall within this category. Until the latter part of the 20th century, such a right was entirely unknown in American law."[58] The logical extension of this reasoning would seem to call into question implicit constitutional recognition of a healthy environment or a stable climate.[59]

When invited to speak to climate change policy, courts thus decline with blushing modesty, insisting that they must defer to the elected branches because it would be undemocratic for unelected judges to weigh in. Courts fail, however, to acknowledge the democratic *harms* of judicial disengagement on climate change policy. The courts' insistence that they have no role in climate change policy perpetuates special-interest-fueled federal climate gridlock that is at odds with popular will and is produced in part by democratic pathologies created or exacerbated by the judiciary itself.

DEMOCRATIC PATHOLOGIES

Clear public majorities support federal mitigation policy in the United States, yet climate change policy gridlock persists. This occurs in part because the constitutional structure of our representative democracy gives outsized political power to less populated states, both in the Senate and in the Electoral College. On key questions—Is global warming caused most by human activities? Should global warming be a high priority for the president and

Congress?—a majority of the public agrees at the national level, but public support falls well below 50 percent in states with a low population but an outsized political voice.[60] The impact of the opposition of low-population, overrepresented states on climate policy is clear. On numerous occasions, the US House has passed significant climate change legislation, only to see it blocked in the Senate.

The impact of the overrepresentation of minority climate change policy views through the Electoral College is also clear. President George W. Bush lost the popular vote when first elected and assumed the presidency over climate advocate Al Gore. Despite promising during his campaign to cut emissions from power plants, President Bush proceeded, in office, to fight tooth and nail against the adoption of federal climate legislation and the use of existing authority under the Clean Air Act to limit emissions. President Donald J. Trump likewise lost the popular vote and then used the full scope of his executive authority to push against federal mitigation policy, including withdrawing the United States from the Paris Agreement. The way that the Electoral College privileges minority climate change policy preferences is also visible in the courts. Of nine justices on the Supreme Court, three (Justices Barrett, Gorsuch, and Kavanaugh) were appointed by a president who lost the popular vote, Donald J. Trump, and two (Justices Roberts and Alito) were appointed by a president who lost the popular when first elected, George W. Bush. And it is those five justices, along with Justice Thomas, who, focusing on concerns about economic impacts on the coal industry and the political controversy surrounding climate change, struck down EPA's effort to use authority under the Clean Air Act to compel a shift away from the coal-fired plants that continue to spew high volumes of carbon dioxide into the atmosphere.

The constitutional structure of our representative democracy thus creates an obstacle to mitigation. While courts withdraw from engaging on climate change policy, reasoning that it would be counter-majoritarian for judges to weigh in, the reality is that the political process that produces climate change process is not, by design or in practice, majoritarian.[61] And in the context of climate change, the Senate and Electoral College resolutely skew against climate change mitigation, allowing a public and political minority to obstruct the adoption and implementation of robust mitigation law.

Moreover, it is not simply that courts decline to exercise their constitutional authority to engage with climate policy out of a mistaken belief that it would be "undemocratic." Courts affirmatively exacerbate the obstacle that our constitutional structure poses to the adoption of mitigation law through interpretations of the Constitution that limit voting rights (thereby decreasing majority voice) and elevate the rights of corporations (thereby increasing corporate voice). Judicial interpretations of the Constitution help to produce, sustain, and afford veto power to the minority public and political block obstructing federal mitigation policy. Fossil fuel interests, gifted a First Amendment right by the Supreme Court to make effectively unlimited donations to political campaigns in *Citizens United v. Federal Election Commission*,[62] orchestrated a well-documented and effective campaign to convince a relevant swath of the American public and their representatives—the group with a smaller population but greater representation through the Electoral College and Senate—that climate change is not real, not human caused, not dangerous, and or/not imminent.[63] And now, in lawsuits brought by plaintiffs against fossil fuel companies seeking to hold them accountable for their lies, the companies, without shame, respond essentially that they have a right to lie under the First Amendment, especially if they are lying to influence public policy. And the constitutional status quo, defined by a series of judicial interpretations and decisions, fails on climate because it fails more broadly on democracy, neglecting to protect fundamental rights like the right to vote while enshrining new rights for corporate interests.

CONCLUSION

The Constitution as presently worded, interpreted, and applied is obstructing the development of a robust societal response to climate change, in part by failing adequately to protect healthy democratic processes and advance real social justice.[64] We could, of course, amend the Constitution to explicitly support protection of the environment and/or better protect democratic processes and advance social justice. We could also encourage new understandings of existing constitutional text that cognize the fundamental value of and right to a healthful environment, more effectively support healthy democratic processes, and go further to advance social justice. Any approach will require an open-eyed reckoning with how and why the constitutional status quo is failing to meet the climate moment.

11 WHAT WE DON'T EXPECT, WHAT WE KNOW BUT IGNORE, WHAT WE SHOULDN'T ASSUME, AND WHAT WE CAN DO

Stan Cox

Climate models tell us how much greenhouse warming the Earth can expect, given various degrees of effort at mitigation. But we know that predictions of both the extent of global temperature rise and its consequences come with large uncertainties. Faced with a future full of what the late former Secretary of Defense Donald Rumsfeld would have called "known unknowns," we nevertheless can be certain of at least one thing: the wise course is to err on the side of caution and reduce greenhouse gas emissions as quickly as possible.

In US politics and governance today, there is much less consciousness of uncertainty than was found in the past. Although there's a recognition that the future is never entirely predictable, it is generally assumed that unexpected occurrences will lie within a fairly narrow range of possibilities. The outlandish trajectory of the twenty-first century so far should have set us straight on that illusion, but it seems that our republic is still groping its way, wholly unprepared, into a dimly lit future filled with Rumsfeldian unknown unknowns.

Given current instability in the spheres of politics and climate, we can expect that hazards—of what kinds and from what sources we don't yet know—will be coming at us even thicker and faster in coming decades. Time and again, unexpected shocks have exposed the fragility of the technologically sophisticated, elaborately integrated societies that the fossil fuel bonanza of the past century has made possible. Over the past half century in particular, sharp rises in the price of a single commodity, oil, have repeatedly caused chaos. The US food system was devastated within weeks of the arrival of the COVID-19 virus, and almost all other basic necessities were eventually hit hard. Current movements to establish antidemocratic ethno-states around the world, including in the United States, have made frighteningly long strides and now threaten to generate cascades of fresh dangers. How can governments and civil society navigate this political/ecological minefield and sustain prospects for a livable future? In attempting to answer that question, it's important not only to admit that there's much we don't know; we should also be critically conscious of what we assume, what we expect, and what we really do know.

ASSUMING WHAT WE DON'T REALLY KNOW

In his 2018 book *The Road to Unfreedom*, Timothy Snyder wrote that the United States had been guided through the 1990s into the 2000s by the "politics of inevitability," which he described as "a sense that the future is just more of the present, that the laws of progress are known, that there are no alternatives, and therefore nothing really to be done." The underlying logic was simple: "Nature brought the market, which brought democracy, which brought happiness." The loss of such confidence, writes Snyder, usually "ushers in another experience of time: the politics of eternity." In this worldview, the nation is caught in a ceaseless "cyclical story of victimhood." Politicians in power "instruct their citizens to experience elation and outrage at short intervals, drowning the future in the present." The politics of eternity aims to rid the future of surprises by forcing society to replicate a carefully constructed past. The country's white nationalists and MAGA politicians inhabit this cycle of resentment, in which a fictional history trumps both the present that we know and the future that we don't yet know.

The politics of both eternity and inevitability are alive and well in the climate policy discussions of the 2020s. The eternalists, for example, foresee the one-time fossil fuel bonanza of the past century being somehow extended into a timeless future. In her influential 2018 article "Petro-Masculinity: Fossil Fuels and Authoritarian Desire," Cara Daggett of Virginia Tech University wrote that the eternalists' political slogan "Make America Great Again" harks back to a time when "achievement of the mid-20th century patriarchal ideal in the US was predicated upon an ongoing supply of cheap fossil fuels. Cars, suburbs, and the nuclear family, oriented around white male workers, formed a triumvirate that yoked the desires of Americans not only to wage labor, but to the continued supply of cheap energy that made the dream possible."

In contrast to the politics of eternity and its straight-ahead disdain for climate mitigation, the politics of inevitability is the source of the technical and economic assumptions on which most mainstream proposals for mitigation are based. Because the politics of inevitability—and what we might call the science and economics of inevitability as well—assume that the future will unfold in familiar ways, these paradigms are not well equipped to address climate in the face of many unknowns that we know lie ahead.

The liberal approach to climate is rooted in an assumption that technological progress is ineluctable. In a 2017 opinion piece for the journal *Science*, titled "The Irreversible Momentum of Clean Energy," Barack Obama urged people not to allow Donald Trump's inauguration to rob them of their optimism about climate. "Businesses are coming to the conclusion," he wrote, "that reducing emissions is not just good for the environment—it can also boost bottom lines, cut costs for consumers, and deliver returns for shareholders." He viewed success on climate as inevitable because "technology advances and market forces will continue to drive renewable deployment"; despite political setbacks, we could rest assured that a red, white, blue, and much greener future lay ahead.

Obama's optimism was reflected in academic papers purporting to show how the United States or world can eliminate greenhouse gas emissions and meet all future energy demand by aggressively developing various combinations of wind, solar, geothermal, hydroelectric, and/or bioenergy capacity. These prospective analyses lay in the realm of what Wes Jackson, president emeritus of the Land Institute, has called "technological fundamentalism." Clearly, alternative sources of energy will be required as fossil fuels are phased out, but the logic of that obvious necessity should not be expected to work in reverse. Building new energy capacity is not sufficient on its own to ensure that oil and gas stay in the ground; more direct policies and respect for ecological limits are needed.

In a 2021 paper in the journal *Nature Communications*, Lorenz Keyßer and Manfred Lenzen evaluated "green growth" scenarios for holding global temperature to 1.5°Celsius above preindustrial levels. These scenarios have been evaluated by the Intergovernmental Panel on Climate Change (IPCC), and all of them assume large increases in the growth rate of wind and solar power capacity, energy efficiency, and/or carbon dioxide capture. But the risk of failing to achieve such historically unprecedented feats in these areas is very high. Keyßer and Lenzen showed that only green *de*growth scenarios—ones not being considered by IPCC—could hold the temperature increase to 1.5°C while assuming realistic increases in efficiency, renewable energy, and carbon capture.

Claims for green growth and the "irreversible momentum" of renewable energy are based on some broad technical and economic assumptions about the near future, some explicit, some unstated. In the following statements, I'll paraphrase some of those assumptions. Relying on them as the foundation for climate policy is, I believe, risky in the extreme:

- Industrial societies will become fully electrified at their current rates of material output; that is, wind-, solar-, and water-generated energy and its storage can and will fill all the roles that fossil fuels now play in transportation, manufacturing, cargo hauling, construction, and agriculture.
- Renewable electric capacity and entire new power grids can be installed—and the built environment and transportation can be remade to run on electricity—at a faster pace than fossil fuels need to be phased out.
- The rate of improvement in energy and material efficiency per unit of GDP produced will consistently outstrip the rate of economic growth far into the future.
- Wind and solar capacity for electricity generation and storage on a scale sufficient to perform all of society's current and growing functions can be developed and deployed without doing unacceptable ecological and human harm through land occupation, ecosystem disruption, economic exploitation, resource extraction, and, potentially, military conflict over resources and territory.
- Because resources are finite, and all energy infrastructure must be replaced repeatedly over time into the long future, materials used in

energy systems' manufacture and installation can and will be completely and repeatedly recycled.

- Over their lifetimes, wind and solar installations will consistently generate five to ten times as much energy as is required to manufacture, transport, install, and recycle them.
- Renewable electric capacity will displace fossil-fuel-generated energy rather than simply adding to the total energy supply.

Researchers have pointed out deep flaws in all these assumptions, and in some cases their physical impossibility. Furthermore, these statements implicitly rest on the broader fundamental assumption that meticulously constructed technological scenarios are invulnerable to unforeseen changes in political, economic, and environmental circumstances—that is, they assume that we can repeal Murphy's Law. For policy-makers to wager that these expectations will all be entirely fulfilled, and therefore fail to pursue bolder policies, would be fatally irresponsible.

"THE ROSE-TINTED LONG RUN"

Technology-led climate strategies foresee more benign energy sources being dropped into the world of industry, where they will be smoothly accommodated; affluent societies can carry on driving, flying, consuming, and growing without pause. A parallel philosophy of inevitability prevails in the world of climate economics, where modelers purport to show that emissions can be sufficiently suppressed by dropping a carbon price into the existing economy without disrupting business as usual. Their goal is to match today's spending on climate mitigation to the expected costs of dealing with the future climate impacts that would occur in the absence of mitigation. For decades, this kind of analysis has been at the heart of debates over climate policy, creating, in the words of Geoff Mann, coauthor (with Joel Wainwright) of *Climate Leviathan: A Political Theory of Our Planetary Future*, "a climate economics for the rose-tinted long run." The models, Mann notes, account for a very narrow range of futures in which political and economic systems remain much as they are today while global temperatures rise gradually and predictably.

The cost-benefit modelers make various choices on behalf of humanity—for example, how much less we should value the well-being of future generations compared with our own—and try them out in their simulation runs. They almost universally assume the indefinite continuance of economic growth, varying only its rate. And they generally assume that faster growth will always make future societies more resilient to climatic chaos. From these and other starting points, the models predict trajectories for the "social cost of carbon" (the optimum price per ton of carbon dioxide to be paid at each point in time to prevent an equivalent amount of future damage).

Modelers tend to conclude that optimized carbon pricing will deliver disturbingly large increases in global temperatures. The Nobel Prize–winning economist William Nordhaus, a pioneer of such analyses, has concluded that even the kinds of policies that would limit warming only weakly—allowing a dangerous 2.5°C increase over preindustrial levels—would be too strict economically; in his words, such policies would "strain credulity." Requiring reductions in greenhouse gas emissions sufficiently steep to prevent catastrophe would be, he wrote, "unrealistically ambitious." The more rational goal, Nordhaus has estimated, is a cataclysmic temperature rise of 3.5°C.

With the world on track to experience that sort of heating later this century, the International Monetary Fund has projected that "the risk of catastrophic and irreversible disaster is rising, implying potentially infinite costs of unmitigated climate change, including, in the extreme, human extinction." To be sure, experts broadly agree that human extinction, while eventually inevitable, is probably not going to happen anytime soon. But in the vast space between a future in which the world economy can carry on growing without interruption and one in which *Homo sapiens* goes extinct within the next century or two, there are countless possibilities, many of them grim. Over the decades, IPCC scientists have attempted to track the risks associated with a host of possible outcomes that could lurk in that space. When in 2022 they issued their most hair-raising release yet, the Sixth Assessment Report, it became chillingly clear that if the global economy stays on its current business-as-usual path, the Earth will become a very different place by the year 2100, and much less livable for human beings.

If humanity fails to rein in emissions quickly and tightly enough, IPCC projected, one-half to three-fourths of the human population could routinely be exposed to "life-threatening" heat and humidity. Food-production systems will be severely undermined. Increased heat stress, drought, soil degradation, destruction of crops by diseases and insects, and extreme weather events could render about one-third of currently suitable cropland unsuitable for farming by 2100. "Multi-breadbasket" crop failures, spanning several world regions at once, would become routine. The number of people at high risk of "hunger, malnutrition and diet-related mortality" would grow by as many as eighty million.

Climate played a role in the mass migrations out of Syria, Central America, and other regions that have been seen over the past decade, and IPCC says that's only a small beginning. By later this century, as many as 3.5 billion people could be compelled to migrate out of their region, nation, or continent by flooding, storms, fires, and/or extreme heat and humidity.

Humanity will be far from alone in its suffering. If atmospheric heating goes unchecked, few species on Earth will be left unscathed. IPCC warns, "In general, adaptation measures can substantially reduce the adverse impacts" of a one- or two-degree rise of global temperature, but beyond that limit, "losses will increase, including species extinctions and changes, such as major biome shifts, which cannot be reversed on human timescales." Do we really want to roll the dice on that?

The world scientific community's projections are all pointing to one conclusion: if, at this late date, governments finally decide to drive down greenhouse gas emissions at the necessary pace, they will have only one shot at preserving a livable future. Therefore, they had better move quickly, boldly, and immediately. An indirect, gradualist approach might have been sufficient back in, say, 1990, but not now. If they do take the politically palatable technology-and-markets route, they will not be able to start assessing progress until after a lag period during which renewable capacity is being built up and carbon prices are being ratcheted upward year by year. That could leave us waking up in the 2030s to the realization that those efforts were not enough—and also to the need for an impossibly precipitous rate of emissions reduction, one much more rapid than the already daunting one we face today. We will have run out the clock on ourselves.

POLICIES FOR A NON-PREDICTABLE FUTURE

The Intergovernmental Panel on Climate Change is saying unequivocally that to keep warming within tolerable limits, global greenhouse emissions must be cut in half by the end of the 2020s. Any policy aiming to accomplish that must be direct and foolproof. So if governments were to come up with climate policies rooted in what we know, rather than what they want to be true, what would those policies look like?

My colleague Larry Edwards and I have been urging that the US government must place statutory caps on the number of barrels of oil, cubic feet of gas, and tons of coal allowed out of the ground annually, with the caps ratcheted down swiftly year by year. To pursue this direct, steep reduction of fossil fuel use would be a bold political decision indeed. But a decision to take a gradual, don't-rock-the-boat approach to climate would be just as deeply political. Nationalizing the fossil fuel industries and imposing declining caps on their output, while very difficult to enact without sweeping political change, could ensure sufficiently deep emissions reductions. In contrast, carbon pricing cannot guarantee that the Earth will avoid a three- or four- or five-degree broiling, and it would be politically toxic to boot. A wide-open public discussion of the contrast between capping and taxing as the basis for emissions reduction might go a long way toward changing the political calculus.

In chapter 9, William Becker describes sixteen ways the US political system can be reformed to bring sufficiently bold climate policies into the realm of possibility. All of these badly needed reforms—among them, balancing the Supreme Court, flushing big money out of electoral politics, ditching the filibuster, quelling tribalism, and fostering a national vision—will themselves be exceedingly difficult to achieve as long as we stay on our current antidemocratic trajectory. But that is no reason to give up on them. As the historian Aviva Chomsky told me recently, regarding climate and political action, "I think we have no choice but to push harder despite everything, on two grounds. One, because even if it seems impossible, we're making it impossible if we don't do anything. And two, because we just have to. Even if there's no hope of success, we still have to, if we're to live with ourselves."

One way to gain greater political acceptance for dramatic climate action is to yoke it to policies that not only soften the direct impacts of greenhouse warming but also reckon with the unintended economic impacts of serious climate mitigation itself. A rapid phase-out of oil, gas, and coal would of course create an urgent need for accelerated renewable energy development; however, we cannot assume that even the most ambitious buildup can fully compensate for the energy gap created by a rapid fuel phase-out. Slowing the phase-out to allow time for renewable energy to catch up would be self-defeating. Therefore, to ensure sufficient, equitable access to energy and other resources, the phase-out of fossil fuels should be accompanied by equally dramatic adaptation policies, and the economy should be transformed in order to function on far less energy.

Through legal tools such as the Defense Production Act, the federal government could respond to diminished fuel supplies by directing energy and other resources toward the production of essential goods and services and away from wasteful and superfluous production. Such policies could include, for example, shifting resources toward renewable energy development instead of military production and prioritizing public transportation over production of private vehicles. In the retail energy sector, price controls, fair-shares rationing of fuels and electricity, and provision of universal basic services could ensure that all households, regardless of income, have access to sufficient energy and other essential goods and services.

A decade ago, the British government studied an energy-rationing strategy similar to what I've just described and concluded that it had "potential to engage individuals in taking action to combat climate change, but is essentially ahead of its time." In retrospect, the proposal was not ahead of its time at all; in fact, it was proposed at what would have been exactly the right time for it to launch. I base that judgment on a UN Environment Program report estimating that if efforts to limit warming to 1.5°C had been launched back in 2010—around the time that the UK proposal was rejected—the world would have needed to cut emissions by a sustained 2 percent per year. That would have been difficult but achievable. But by 2021, when the UN report was published, the required rate of emissions reduction had ballooned from 2 percent to almost 8 percent per year. At this late date, no climate policy, no matter how bold, could possibly be considered ahead of its time.

Over the course of that lost decade, I became more and more pessimistic about political prospects for climate action. The election of Donald Trump guaranteed that the United States would reach the year 2021 still with no national climate policy at all—forget about a declining cap on fossil fuels, industrial planning, and rationing. Trump is long gone from the White House, but those policies remain beyond our reach. The passage of the 2022 Inflation Reduction Act with its climate provisions was welcome, but, rooted as it is in the politics and economics of inevitability, that law is only a very small step in the right direction. It also risks creating a sense of complacency: "OK, we took care of climate. What's our next issue?"

Meanwhile, despite the setbacks it suffered in the 2022 elections, the racist, antidemocratic politics of eternity have not receded. Our predicament today could not be clearer. The struggle to crush greenhouse gas emissions is no longer separable from the country's struggle to prevent a decline into autocracy and instead become, for the first time, a multiracial pluralistic democracy. Building pro-Earth congressional and state legislative majorities is essential to taking down fossil fuels, reforming agriculture, and so on. But doing that entails a rearguard action to stamp out, permanently, all attempts by far-right extremists to sieze and keep unified control over federal and state governments deep into the future.

NAVIGATING THE UNKNOWN UNKNOWNS

Whatever comes of this struggle over the very nature of the United States, communities will need to prepare for life in a much hotter world. But with the realms of politics and weather each becoming increasingly unstable and unpredictable, it seems harder than ever to even figure out what to prepare for. This brings us to the quest for a kind of general-purpose adaptive capacity—in policy shorthand, resilience. The term has long been wending its way through research, policy-making, and everyday language, taking on new meanings as it goes—or often simply being left undefined. The concept of resilience emerged first in the discipline of psychology during the 1950s. Two decades later, it was adopted by systems ecologists, who defined it as an intrinsic property of ecosystems. But since about 2000, the policy world has most often applied the term to human communities, often in the context of adaptation to climate change or natural disasters in general. These days, resilience slips back and forth between signifying communities' or societies' inherent capacity to endure hardship and bounce back or, alternatively, a goal toward which they must aspire if they are to navigate a world full of hazards both known and unknown.

It is widely assumed that affluent societies can achieve resilience by amassing sufficient material resources to ride out whatever adversities the world throws at them, then recover, build back, and reduce their vulnerability to future calamities. In this view, resilience is mostly a by-product of economic growth. So where does that leave societies or communities that are vulnerable to climatic disasters but have no possible path to getting rich? The answer that the have-nots typically hear from the haves at global climate summits is that they must be super-resilient, achieving through sheer pluck and ingenuity the capacity to absorb unforeseen shocks and endure the inexorable, all-too-predictable heating of their world. If that's the case, the world's vulnerable nations insist, then the United States and other nations of the Global North must provide plentiful funding for the climate loss-and-damage fund to which the world's nations agreed at the COP27 climate summit in November of 2022. The international groups Climate Action Fund and Feasta have published proposals for well-designed systems that would incorporate that kind of redistributive climate justice into agreements for rapidly reducing greenhouse gas emissions globally.

The United States' fair share of payments owed to the Global South for climate mitigation and adaptation, plus loss-and-damage reparations, could amount to $160 billion per year over the next decade (as estimated by Climate Action Fund). Washington can easily afford to pay its $1.6 trillion climate debt. The Pentagon could provide the money, for instance, by simply calling off production of the $1.7 trillion F-35 fighter jet and divert the savings toward the global fund.

Meanwhile, in the Global North, resilience-through-wealth-accumulation must give way to resilience-through-ecological limits (a shorter way of saying "resilience-through-ceasing-production-of-the-climate-chaos-to-which-the-world-is-trying-to-be-resilient"). These requisites have been obvious for a long time, going back at least to the 1990s and the concept of "contraction and convergence," under which affluent countries would deeply reduce their greenhouse gas emissions, converging with low-income nations who need access to more energy and other resources, even if it means increased emissions for some time.

Within the United States, a cap on fossil fuels with planned resource allocation, price controls, rationing, and universal basic services would be the domestic equivalent of contraction and convergence. Although I expect that economic modelers would view those measures as "unrealistically ambitious," the barriers to implementing them are not physical but rather political and economic, and they can be overcome. We know that such policies are not intrinsically unachievable, because the United States has deployed them before, in the civilian economy of the 1940s. Fuels and other resources were capped, partly because of import cutoffs and partly by the diversion of resources to the war effort. The War Production Board was set up to steer scarce resources toward essential production and halt other production. In the consumer economy, scarcity and inflation prompted the development of extensive, locally administered systems for price controls and rationing. Policies like these and others that have been proven workable should not be struck from the list of the possible only because they cannot be passed into law under current political conditions—especially when the broader goal is to change and transcend those very conditions.

We can't afford to give up on the quest for national and international climate mitigation policies, especially the fast phase-out of fossil fuels. But with the future becoming less predictable year by year, it would be foolhardy to assume that such an opening will come soon. With the required rate of emissions reduction having quadrupled in just the past ten years, another decade of delay, or even a few years, would leave the world with a climate budget so small that we would inevitably blow through safe limits. However, safety and danger are not absolutes but matters of degree—in the case of climate, literally. Every tenth of a degree of warming that's prevented will preserve some measure of livability; therefore, the more mitigation the better, even if it's not as much as we are aiming for. Conversely, every year of delay in climate mitigation, especially a continued procrastination in curbing fossil fuel extraction, will trigger a need for adaptation to increasingly harsher climatic conditions, and perhaps impossibly deep reserves of resilience. We must be ready to take advantage of any political window, no matter how small, even it doesn't open up until years from now.

SOME THINGS WE KNOW THAT WE DO KNOW

Whatever the political forecast in Washington, we can expect local, state, and regional activism to become more important than ever in fighting fossil capitalism and achieving environmental justice in the United States. We could, one hopes, see a surge of participation in movements like Extinction Rebellion and youth climate strikes, as well as others that are being discussed: labor actions, consumption strikes and boycotts, and even the kind of general strike advocated by the group Earth Strike.

The environmental justice movement, led by Black, Latino, and Indigenous communities, provides many other examples. Across decades, environmental justice groups have won improvements in community health and quality of life by fighting the electric utilities, fuel refiners, cargo transporters, and other dirty industries that ruin local air and water. The long and often successful struggle against pipelines, power plants, and other fossil infrastructure on Native lands has provided inspiring examples of how to curb climate change and also protect communities and ecosystems against the abuses that cause climate change. These and other grassroots efforts can make a big difference. In 2021, the Indigenous Environmental Network and Oil Change International reported that successful or ongoing Indigenous-led struggles against fossil fuel infrastructure across North America had the potential to make a whopping big impact on climate change. If all of these struggles against fossil fuel infrastructure ultimately were to prevail, the groups estimated, they could prevent greenhouse gas emissions equivalent to almost one-fourth of North America's total annual emissions—as much as are released by four hundred coal-fired power plants or more than 340 million passenger vehicles in a year.

As important as grassroots action is for mitigation, it can't eliminate emissions quickly enough unless it is backed up by strict federal enforcement of limits on fossil fuel use and other threats to humanity and the Earth. If no such backup is forthcoming, emissions can still be reduced and worse-case scenarios avoided, but the Earth's climate will become more and more chaotic. We and future generations will face the need for profound adaptation.

With few prospects for strong national policies, in the present political climate, regional and local adaptation will become more crucial than ever. This is an area in which we can dare to hope for positive surprises. Examples of equitable, small-d democratic governance of local resources are few and far between today, but they exist and are being discussed more and more. Especially if we're saddled with a federal government that is taking a position of neglect, whether benign or malign, toward the majority of Americans who want justice for humans and the Earth, we will need more grassroots democracy. William Becker suggests one much-needed reform: a constitutional amendment enabling processes for direct democracy. The most well-known such mechanism, the ballot initiative, can serve democratic ends even when a measure is defeated. In 2022, the voters of deep-red Kansas and Kentucky transformed the national struggle for reproductive rights, and much more, when they decisively voted down state constitutional amendments that would have deprived women of their reproductive rights. We need a flowering of other forms of human-scale democracy as well: citizens' assemblies, worker-owned cooperatives of all kinds, mutual-aid movements, participatory budgeting of public funds, citizens' juries, and other varieties of small-to-medium-scale deliberative democracy yet to be conceived.

We don't know how much of this will happen, but there's one thing we know for certain. Protecting voting rights and reforming the electoral process today, in order to keep the hands of the coup-plotters and would-be autocrats off the levers of power tomorrow, will greatly increase the probability that we can achieve, for the first time, a multiracial, pluralistic democracy. That, in turn, can revive prospects for a livable future.

IV EDUCATION FOR UNCOMMON SENSE

12 ACADEMIC CULTURE, DEMOCRACY, AND CLIMATE CHANGE

Michael M. Crow and William B. Dabars

It is sobering to realize that the 2,500-year trajectory of the academy has culminated in the dysfunction of our democratic institutions and a lagging response to global climate change. Despite incalculable contributions to human well-being, academic culture is implicated in both social discord and our failed relationship with the natural systems on which we depend. Whether we loosely characterize its modus operandi as the technological domination of nature or attribute its skew to the legacy of Cartesian dualism, the scientific culture at the heart of the academy perpetuates the fallacy that it is reasonable to analyze humanity in isolation from nature. By partitioning academic disciplines and valorizing increasingly specialized or reductionist knowledge motivated primarily by curiosity rather than the collaborative pursuit of social goals that equitably meet the basic needs of our fellow citizens, academic culture has abandoned its best hope to make sense of the holistic interdependence of humankind and the complex biogeochemical cycles that constitute the Earth system.

The complexity of the challenges evident at the intersection of democratic dysfunction and environmental destabilization requires moving beyond the limitations of our present academic culture, which has informed the American experiment in democracy since the seventeenth century as well as presiding over the scientific research and technological innovation that has sometimes delivered inadequate environmental outcomes. But any effort to analyze or ameliorate the disequilibrium of the environment in isolation from social, economic, cultural, and political contingencies will inevitably fall short. Given the urgency of the current polarized political context, however, it is imperative that the academy reexamine its role in preserving our constitutional democracy.

Entirely apart from issues associated with democratic governance and climate change, the shortcomings of contemporary research universities are well-known. The exclusion of academically qualified applicants from our leading colleges and universities, for example, is a consequence of admissions protocols that reward privileged students. The prioritization of basic research over application and outcomes, and individual attainment over transdisciplinary collaboration, diminishes the social impact of knowledge production. Building on the patterns and practices institutionalized in research universities in the nineteenth century, academic culture must recognize, acknowledge, and overcome its limitations as well as the contingency and unpredictably of the wicked problems presented by the intersection of democratic dysfunction and climate change.

Academic culture in the United States has long invoked tenets of social responsibility founded on democratic values while simultaneously facilitating research that has degraded the environment and subverted the equitable distribution of the benefits of science. Therefore, any effort to reinvigorate and revitalize our experiment in democracy requires that our nation's colleges and universities differentially affirm fundamental responsibility for the social, economic, cultural, political, scientific, and technological well-being of their local communities. In addition, universities need to restructure their epistemic, administrative, and social foundations to operationalize transdisciplinary research that promotes collaborative engagement by focusing on the outcomes of knowledge production.

The academy in the United States, which emerged in the wake of the scientific revolution and the Enlightenment and evolved alongside the gathering momentum of the Industrial Revolution, has contributed preponderantly to the epistemic base of scientific knowledge and technological innovation; however, it was never designed to guide society through the rapid changes triggered by the accelerating pace of modernity. Steven Pinker identifies reason, science, humanism, and progress as the interrelated ideals of that "great Enlightenment experiment, American constitutional democracy with its checks on government power."[1] But the disappointing outcomes of efforts to reconcile representative democracy with the responsible stewardship of the planet should come as no surprise.[2] Although the philosophical underpinnings of our democratic experiment were pragmatically balanced by the founders, the pivotal formulations of the US Constitution failed to protect nature. Moreover, the principles of capitalism as articulated by Adam Smith in *The Wealth of Nations* imposed no limits on economic individualism or the inclination of societies to exploit natural resources capriciously.

Whereas Francis Bacon specified the purpose of scientific discovery to be the "relief of man's estate," no account of the increasing threats posed by a destabilizing climate can overlook their genesis in the technological domination of nature undertaken in the name of progress. Or perhaps, more fundamentally, the Cartesian dichotomy encourages the delusion that we may consider humankind in analytical isolation from nature. These inadequate outcomes underscore our contention that organizations such as research universities, which are committed to discovery and innovation, must recognize that knowledge production and technological innovation are not automatically aligned with beneficial overarching social goals.[3] Instead, comprehensive and sustained efforts will be required to integrate the quest to advance discovery, creativity, and innovation with an explicit mandate to assume responsibility for society.[4] Approaches that ameliorate the interrelated conundrums that now plague the Earth's systems will require systems-level thinking that challenges the reductionist assumptions of the Enlightenment.

Humanity may be defined by its "paleolithic emotions, medieval institutions, and god-like technology," the naturalist and ecological theorist E. O. Wilson once quipped.[5] Among medieval institutions, none has proven to be more enduring than the university, which must now respond to both political dysfunction and anthropogenic interference with natural systems. The innovations needed to mitigate or adapt to climate change must be matched by an assessment of socioeconomic factors and human fallibility. It is, after all, the deficiencies of the democratic process that have allowed the election of unscrupulous politicians who deny climate change or obstruct efforts to combat environmental degradation. Scientific or technological illiteracy among policy-makers and elected officials is matched by a growing affluent class that valorizes individualism over civic engagement and is insulated from complex sociotechnical issues. Such an account should begin with an effort to mediate the tensions between the impacts of economic individualism on the one hand and the ravages of human incursion on natural systems on the other.[6] This approach builds on the systems-level thinking advocated by science and technology policy scholars Brad Allenby and Dan Sarewitz, and the "ecological design intelligence" proposed by David Orr, which, as he explains, refers to our "capacity to understand the ecological context in which humans live, to recognize limits, and to get the scale of things right." Moreover, he continues, "It is the ability to calibrate human purposes and natural constraints and do so with grace and economy."[7]

THE "MAJESTY AND MECHANICS" OF OUR DEMOCRACY

"Universities are shunning their responsibility to democracy," Johns Hopkins University president Ronald J. Daniels contends. He reflects on his surprise and dismay when, as a recent arrival from Canada, he was disappointed in his expectation that the United States would be a "bastion of civic learning." He had thought that "surely the stewards of the world's first modern democracy would understand the need to cultivate an understanding of both its majesty and its mechanics—the Enlightenment ideas that animate it and the institutions that make it work."[8] As he and colleagues put it in a 2021 book, our colleges and universities are "essential to the flourishing of liberal democracy" because they are responsible for defending the "liberal democratic experiment as institutions that enrich and are enriched by democracy, and are inextricably intertwined with democracy's values and ends."[9] The majesty and mechanics of our democratic values and ends were illustrated by the leaders of the founding generation, who, as political theorist Suzanne Mettler observes, "strongly believed that by encouraging and subsidizing advanced learning, the nation would foster the knowledge, creativity, dynamism, leadership, and skills that would spur economic growth, technological innovation, and social advances."[10]

The Constitution of the Commonwealth of Massachusetts, for example, which was drafted by John Adams and served as a prototype for the

Constitution of the United States, refers to the imperative that education be "diffused generally among the body of the people, being necessary for the preservation of their rights and liberties." Adams deemed it the "duty of legislatures and magistrates, in all future periods of this commonwealth, to cherish the interests of literature and the sciences, and all seminaries of them." Inasmuch as the seminary in question was Harvard College, founded in Cambridge, Massachusetts in 1636, Adams specified, "especially the university at Cambridge."[11]

At the Constitutional Convention, in Philadelphia in 1787, James Madison championed the formation of a so-termed national university, which was envisioned as a federally chartered institute for scientific research and scholarship. "The founders were motivated," Albert Castel explains, by their "conviction that our experiment in republican government could not succeed unless the people and their officials were properly educated."[12] Thomas Jefferson promoted a constitutional amendment to establish a national university. After failing to garner sufficient support, he resolved to create the University of Virginia as a surrogate.[13] Jefferson repeatedly expressed the imperative for an educated citizenry. In 1820, for example, he wrote, "I know of no safe depository of the ultimate powers of the society, but the people themselves; and if we think them not enlightened enough to exercise their control with a wholesome discretion, the remedy is not to take it from them, but to inform their discretion by education."[14]

By enacting the Morrill Act in 1862, Congress reified the commitment to democratic values in the United States by establishing a system of colleges and universities that were supported by grants of land owned by the federal government. "Without excluding other scientific and classical studies," land-grant schools were obligated to "teach such branches of learning as are related to agriculture and the mechanic arts" and to expand the "liberal and practical education of the industrial classes."[15] Indeed, the scientific culture and technological innovation institutionalized in land-grant universities contributed to the consolidation of the present model of the research university in the United States during the final quarter of the nineteenth century. This distinct institutional type, epitomized by the founding of Johns Hopkins University in 1876, hybridizes the British and German academic models by integrating undergraduate and graduate education with advanced scientific research.

The correlation between democracy and an educated citizenry was reaffirmed when, in the context of proposing an "economic bill of rights" during his State of the Union address in 1944, President Franklin Roosevelt recommended that the "right to a good education" was an "economic truth" that had become "self-evident."[16] The imperative for the academy to advance civic learning was made explicit by the Commission on Higher Education, which issued a report in 1947 that observed that the "first and most essential charge upon higher education" is to transmit "democratic values, ideals, and processes." Among the overarching goals was the "fuller realization of democracy in every phase of living." Implicit within this goal was that higher education could "liberate and perfect the intrinsic powers of every citizen," which was deemed the "central purpose of democracy." Accordingly, the commission directed the federal government to partner with states to support universities by providing the "means by which every citizen, youth, and adult is enabled and encouraged to carry his education, formal and informal, as far as his native capacities permit." The report presciently warned of the consequences of restricting access to higher education: "If the ladder of educational opportunity rises high at the doors of some youths and scarcely rises at the doors of others, while at the same time formal education is made a prerequisite to occupational and social advance, then education may become the means, not of eliminating race and class distinctions, but of deepening and solidifying them."[17]

THE ACADEMY AND CIVIC LITERACY

For most Americans who attended liberal arts colleges or the undergraduate colleges of research universities, curricula were structured by general education requirements. As Louis Menand explains, general education reflects "what the faculty chooses to require of everyone" as part of an overarching pedagogical philosophy "even when the faculty chooses to require nothing."[18] Surprisingly, the initial impetus came from a federal government program introduced during the final months of World War I. As Daniels and colleagues explain, the US War Department developed a mandatory course, titled War Issues, that introduced students to the justifications for the war. Although thought to be imperfect, most institutions continued the course after the Armistice because it had "answered a felt need among colleges to renew and reimagine higher education's civic purpose through required courses." As an alternative to the unfocused outcomes of the elective system, the course "created an opening for the reintroduction of a core education in democracy back into colleges and universities."[19]

In 1945, Harvard instituted the most influential postwar general education effort with the publication of *General Education in a Free Society*, commonly called the Harvard Red Book.[20] The committee explained that the intent was to describe how general education could be used to "both shape the future and secure the foundations of our free society." Daniels and colleagues elaborate: "The report's authors took the view that as much as democratic citizenship relies on freedom of thought and independent insight, it also demands, to an equal degree, a common set of understandings and capacities."[21]

The process of civic education would ideally begin well before college, as the Harvard classics scholar Danielle Allen and colleagues point out in the recent report *Educating for American Democracy*: "In recent decades, we as a nation have failed to prepare young Americans for self-government, leaving the world's oldest constitutional democracy in grave danger, afflicted by both cynicism and nostalgia, as it approaches its 250th anniversary."[22] Nevertheless, it is essential that colleges and universities inculcate these skills at scales that are sufficient to meet challenges posed by complex problems like climate change as they are coproduced by inconsistent and unpredictable governance structures.

To address the breach between the goals and outcomes of civic education, the Arizona Board of Regents, which governs the three Arizona universities—Arizona State University, Northern Arizona University, and the University of Arizona—is "leading an effort to reimagine the structure and requirements of the Arizona General Education Curriculum."[23] The policy requires the universities to produce graduates who can act as "informed citizens in a robust constitutional democracy based in values of individual freedom, self-reliance, and equality under the law, diversity, inclusion, and constructive dialog through civil discourse."[24]

Accordingly, Arizona State University has proposed to redesign its framework for general education, which, among proposed multiple knowledge areas, includes civic engagement. This objective reflects the overarching commitment of the university to social responsibility, which is confirmed in the university's first charter, adopted in 2014: "Arizona State University is a comprehensive public research university, measured not by whom it excludes, but by whom it includes and how they succeed; advancing research and discovery of public value; and assuming fundamental responsibility for the economic, social, cultural, and overall health of the communities it serves."[25] Along with critical thinking and inquiry, creativity and innovation, and self-reflection and lifelong learning, proposed foundational skills and competencies include communication and civil discourse, and civic and global responsibility. Civic engagement comprises three proposed dimensions that enjoin students to understand US institutions, negotiate competing perspectives, and "demonstrate awareness of cultural differences in relation to political systems, legal structures, history, economy, values, beliefs, and practices."

The mission statement of the Democracy Initiative of the College of Liberal Arts and Sciences, which organized the inaugural Democracy and Climate Change Conference in 2022, expresses a complementary institutional commitment that links the issues that headline this chapter and book:

> A robust democracy is a powerful asset as we address the challenges of climate change. Accordingly, we will seize a historic opportunity to prepare our students to meet the interrelated challenges to democracy and climate stability and to contribute to the transition to a post-fossil fuel world that is democratic, fair, resilient, and durable. Our students will graduate knowing how the earth works as a physical system, the basic principles of democracy, and why both are important for their lives and careers; a new generation of leaders who understand that we are citizens of one indivisible civic and ecological community. Further, we will deploy our assets of teaching, research, expertise, public events, counseling, career planning, and leadership training, and engage our alumni in the effort to preserve a habitable planet and the hard-won rights of people to choose how they are governed, by whom, and to what ends.[26]

Since "good citizenship consists of a multifaceted set of competencies," Ron Daniels calls on colleges and universities to impart "knowledge of democratic history, theory, and practice; skills of reasoning, persuasion, and interaction with political institutions and community organizations; an embrace of core democratic values like tolerance and the dignity of all people; and aspirations toward cooperation and collective action."[27] To promote ecological literacy needed to address climate change, David Orr highlights the "irreducible body of knowledge that all students should know, including how the earth works as a physical system." Accordingly, he calls on colleges and universities to impart "basic knowledge of ecology and thermodynamics, the vital signs of the earth, the essentials of human ecology, the natural history of their own region, and the kinds of knowledge that will enable them to restore natural systems and build ecologically resilient communities and economies."[28]

Although our nation's leading colleges and universities would ideally use their sizeable endowments to explore alternative models and expand access, it is especially important for large public research universities to scale up efforts to counter inequitable admissions practices by allowing all learners as well as society to derive the democratic spillover benefits of higher education. Inasmuch as access to knowledge underpins the social objectives of a pluralistic democracy, accessibility facilitated by scalability must be at the core of evolving institutional models. Educating students from the top 5 or 10 percent of their high school classes represents a de minimis contribution by our leading colleges and universities. Instead, the challenge is to educate to internationally competitive standards of achievement the top quarter or third of successive cohorts of eighteen to twenty-four-year-olds, and for our public universities to provide opportunities for lifelong learning to more than half the population of the United States.[29]

Economists such as Raj Chetty, Claudia Goldin, Lawrence Katz, Walter W. McMahon, and Emmanuel Saez have documented the private and social benefits of higher education.[30] Nevertheless, conversations about equity and excellence must not focus simply on producing more college graduates, because not all bachelor's degrees are equivalent.[31] Mere access to standardized forms of instruction will not deliver desired democratic outcomes. Nor is narrowly focused vocational or technical training sufficient to prepare graduates for future cognitive challenges and workplace complexities needed to address climate change.

Undergraduate education that imparts both civic literacy and ecological intelligence will be based on comprehensive curricula that integrate the liberal arts and sciences with the production of cutting-edge knowledge which will be essential to the workforce of the future global knowledge economy. Arizona State University advances these objectives through two models of the American research university that have been operationalized during the past two decades. First, the New American University model envisioned by ASU president Michael M. Crow demonstrates that major public research universities can differentially manage the tensions between broad accessibility and academic excellence to maximize social impact.[32] Second, the Fifth Wave model extends the objectives of the New American University by envisioning the emergence of a subset, or league, of similarly committed public research universities.[33] Both models democratize American higher education. As we posit in a forthcoming book, large-scale public research universities must differentially accommodate two distinct groups of learners: (1) successive cohorts of traditional on-campus students (eighteen- to twenty-four-year-olds) from increasingly diverse socioeconomic and demographic backgrounds who seek undergraduate degrees within programs based on funded research that is embedded in the liberal arts and sciences; and (2) everyone else, referring to all populations of learners who would benefit from advanced education and training, especially the thirty-six million Americans who have attended college but have not completed their degrees. Of course, this is to say nothing of the critical roles of the research university in discovery and innovation, graduate and professional education, and societal engagement.[34]

To further these ends, a subset of public research universities must assume a broader mandate by redefining academic culture as consisting of a set of differential platforms for universal learning. This would "enable qualified students within their communities, regardless of socioeconomic status or life situation, to acquire the knowledge and skills needed to achieve their goals by empowering them to freely shape their intellectual development and self-determined creative and professional pursuits." A system of higher education that rewards only the privileged few fails to animate hope in meaningful social progress. Rather, higher education must be reconceptualized as an abundant system, which, like languages or open information systems, becomes more valuable for individuals and society when it is broadly adopted and widely used. An abundant system of undergraduate education implemented at national scale would transform our democracy and its ability to ameliorate climate change.[35]

REDRESSING GLOBAL CLIMATE MANAGEMENT FAILURE

As "Earth's unruly tenant," to invoke the eloquent phrase coined by environmental scientist Jane Lubchenco,[36] humanity has unleashed untold harm on the planet, which, as delineated in the classic account of Peter Vitousek and colleagues, includes transforming up to one-half of the Earth's land surface, destroying ecosystems, degrading the atmosphere, exhausting natural resources, and causing rampant extinction of animal species.[37] Assessment reports of the Intergovernmental Panel on Climate Change corroborate the extent to which any prospect for equilibrium between humankind and nature is unraveling.[38] Of course, climate instability also interacts with and often exacerbates other wicked problems that challenge democracies and authoritarian regimes alike, including overpopulation, widespread hunger, and economic insecurity.

By facilitating scientific discovery and technological innovation, academic culture has contributed to the most sweeping and uplifting economic, social, and technological transformation in human history. However, in the estimation of Allenby and Sarewitz, the transformation has become "so technologically and socially complex that the Enlightenment thinking that spawned it may be more harmful than helpful when it comes to guiding our actions."[39] To help understand the scope of the environmental and governance challenges facing us, Allenby and Sarewitz propose a three-level heuristic to describe and assess our interactions with technology. Within this schema technological artifacts (for example, vaccines or nuclear reactors) comprise Level I. Systems or networks of associated technologies that include supporting physical or social infrastructures (for instance, manufacturing or transportation) comprise Level II. "But there is a third level that we are not so familiar with, a level at which technology is best understood as an Earth system—that is, a complex, constantly changing and adapting system in which human, built, and natural elements interact in ways that produce emergent behaviors which may be difficult to perceive, much less understand and manage."[40]

Climate change or political dysfunction, for example, are both Level III challenges. The conditions described by Level III transcend disciplinary conceptualization and are best understood as complex adaptive systems that exhibit emergent, or nonlinear, characteristics and feedbacks that interact in contingent and unpredictable ways. The techno-human condition of Level III does not respond in consistent and comprehensive ways to inquiries or interventions. Instead, responses at this level are incomplete, ambiguous, contradictory, and uncertain. As a result, researchers often commit category errors by making Level III predications based on applying quantifiable results found at Levels I and II to Level III. The resulting forecasts are often subsequently sabotaged by competing nonlinear, emergent properties or feedbacks that lurk within Level III.

Allenby and Sarewitz summarize their recommendations for engaging with contingent and unpredictable Level III crises that are epitomized by simultaneous democratic dysfunction and climate change: "Forget about 'solutions'; expand option spaces; expand the number of voices; make more frequent but smaller decisions; encourage questioning and continual learning, and dialog with Earth systems." For instance, although environmental science dominates research into climate change, the private sector generates the most greenhouse gases and, therefore, has the most knowledge about the options needed to reduce or eliminate emissions. Nevertheless, environmental scientists have typically excluded industry practitioners. Consequently, a narrow band of nongovernmental organizations, activists, and experts appointed by policy-makers have produced top-down control fantasies that fail to fully explore the option spaces. The current process has become imprisoned in responding to historical errors rather than anticipating future possibilities. Therefore, instead of providing insight into future scenarios that are contingent and unpredictable, experts and politicians treat computerized models as if they are windows on the future. To respond to amorphous and changing conditions, political and academic institutions need to recur to the option for negation fostered by the Enlightenment in which "truth" is constantly challenged.[41]

To address Level III challenges associated with efforts to keep our planet habitable and to realize a future in which prosperity and well-being are broadly attainable, ASU has operationalized the Julie Ann Wrigley Global Futures Laboratory (GFL). A large-scale platform for transdisciplinary knowledge production and technological innovation, GFL positions the university as a global hub of networks of scientists and scholars that seek to establish a new equilibrium between humankind and the dynamic Earth system. Among critical issues associated with anthropogenic pressure on the environment, GFL is addressing the depletion of natural resources, degradation of the environment, water scarcity, food security, energy systems, environmental and public health, and governance and policy. At the heart of GFL is the Global Institute of Sustainability and Innovation and the College of Global Futures, which comprises three schools: the School of Sustainability; School for the Future of Innovation in Society; and School for Complex Adaptive Systems. Associated units include the Consortium for Science, Policy, and Outcomes, which seeks to enhance the contributions of science and technology to an improved quality of life, with particular attention paid to questions of who is likely to benefit or suffer from public investments in knowledge production and innovation. Consistent with its ameliorative mission, GFL is housed in the new Rob and Melani Walton Center for Planetary Health at ASU.

As the nation's largest university governed by a single administration, ASU is committed to leading efforts to institutionalize sustainability practices by example. ASU has undertaken a broad array of initiatives aimed at making resource systems circular and impacts on the climate positive, promoting collaborative action leading to community advancement, optimizing appropriate uses of water, engaging the academic community in sustainable action, and promoting resilience to the threats posed by climate change. Sustainability practices at ASU conserve and monitor energy consumption, promote renewable energy such as wind power, reduce solid waste, improve the quality of food, provide healthful water, enhance sustainable transportation options, reduce the cost of buildings and grounds, and improve services, maintenance, and purchasing. ASU monitors and reports on the status of these efforts by submitting data to external monitoring programs such as STARS, the Sustainability Tracking, Assessment, and Reporting System of the Association for the Advancement of Sustainability in Higher Education.

Dealing with the pernicious impacts of climate change and promoting more sustainable alternatives will require unprecedented innovation at appropriate scales. The long-standing commitment of academic culture to social responsibility notwithstanding, novel approaches are required to institutionalize responsible innovation along interrelated dimensions of anticipation, inclusion, reflexivity, and responsiveness.[42] Accordingly, ASU adheres to these four dimensions on a de facto basis. For instance, ASU established the Decision Center for a Desert City (DCDC) in 2004, which coproduces climate, water, and decision research to bridge the boundaries between scientists and policy-makers as a university-based boundary organization.[43] DCDC helps researchers *anticipate* ways to address the megadrought that continues to plague the Colorado River Basin. Instead of producing idealized, disembodied scenarios, DCDC allows researchers to engage with communities by producing socially robust scenarios. ASU has demonstrated its commitment to *inclusion* by increasing access to members of underserved communities pursuant to the terms of its charter. ASU puts *reflexivity* into practice with the Socio-Technical Integration Research, or STIR, program, which helps scientists assess the potential consequences that may result from their innovations. Finally, by explicitly assuming responsibility for the communities it serves, ASU affirms its commitment to *responsiveness*.[44] These interdependent dimensions of responsible innovation continue to help ASU coproduce responses needed to address the complex Level III problems posed by climate change as well as help society resurrect the promise of democracy.

TOWARD A NEW ACADEMIC CULTURE

The pragmatist philosopher John Dewey argued more than a century ago that "democracy must be born anew every generation, and education is its midwife."[45] He observed that social progress starts with the "problem of discovering the needs and capacities of collective human nature" and proceeds by then "inventing the social machinery which will set available powers operating for the satisfaction of those needs."[46] As organizational theorists Ann Pendleton-Jullian and John Seely Brown put it, society must undertake "unbound design" in a "white water world," which is "rapidly changing, increasingly interconnected, and where, because of this increasing interconnectivity, everything is more contingent on everything else happening around it."[47]

As philosopher Hans Jonas observed, the magnitude of the "Promethean" technological accomplishments of contemporary society moves responsibility with "no less than man's fate for its object, into the center of the ethical stage." (Consistent with the themes of this book we note that etymologically, Prometheus represents forethought. In contrast, his mythological brother, Epimetheus, epitomizes our infallible but unresponsive 20/20 hindsight.) Responsibility, according to Jonas, is a "correlate of power and must be commensurate with the latter's scope and that of its exercise."[48] Of course, reason itself has increasingly come under assault and science under siege. As political theorist Yaron Ezrahi has observed, "Contemporary democracy is not the deliberative self-governing polity of informed free citizens envisioned by modern Enlightenment thinkers."[49]

Addressing Level III challenges such as the preservation of our democracy amid the emerging global crisis of rapid climate change requires that we recalibrate our academic culture. Organizational change may be construed as the synthesis of evolutionary processes and deliberate intervention, which could be termed institutional design. Herbert Simon examined the distinction between the natural sciences and what he famously termed the "sciences of the artificial" half a century ago, but that convenient binary is insufficient when applied to Level III wicked problems like climate change that operate on all levels simultaneously.[50] Although we may be "empowered by the tools of the Enlightenment," as MIT computer scientist Danny Hillis observes, in this postmodern era we seem to have entered what he terms as the "Age of Entanglement": "We can no longer see ourselves as separate from the natural world—or our technology—but as a part of them, integrated, codependent, and entangled."[51] We concur with Allenby and Sarewitz that the mismatch between our accustomed reductionist Enlightenment thinking and what is required to understand the complexity of the Earth system demands "nothing less than a new frame of reference for understanding and for action: a reinvention of the Enlightenment." Or, put another way: "Either we accept that we are impotent brutes living way beyond our means because of the technological house of cards we occupy or we search for a different set of links to connect our highest ideals to the reality we keep reconstructing."[52]

Although we herald the arrival of the knowledge economy, we also need to acknowledge our ignorance so that we can motivate ourselves to find new ways to pursue innovations that will mitigate or adapt to climate change as well as reconstruct supportive political institutions. Hubris prevents us from seeing the contingent and unpredictable nature of Level III systems. For instance, democracies continue to fund complicated computer models that aim at predicting the nature of climate change in futile attempts to control nature instead of expanding the capacity of democracies to deal with alternative scenarios. If universities are to participate in managing the contingent, unpredictable, and emerging Level III challenges posed by climate change and the governance crises that gnaw at the legitimacy of democratic institutions, they must recalibrate academic culture so that it promotes pragmatic and concerted local actions at scales sufficient to be effective.

At the same time, we fail to live in harmony within our contingent, unpredictable, and dynamic techno-human condition. Sustainable development, adaptive management, industrial ecology, and intergenerational equity can serve as organizing principles for the production and application of socially robust knowledge in a reconceptualized academic culture. Inherent in all these concepts are the flexibility, resilience, and responsiveness that needs to be built into the academy as well as its interrelationship with business and industry, government agencies and laboratories, and organizations in civil society. While acknowledging our ignorance, our research universities need to devise qualitative metrics to assess efforts aimed at reducing uncertainty by pragmatically acting and observing outcomes. Such experiments will help us to define threshold criteria for responsive scenarios that can sustainably adapt to emerging risks. Instead of invoking the precautionary principle, which paralyzes needed innovation and beneficial risk-taking, we must increase organizational and institutional innovation as well as come to terms with our fallibility and limitation.[53] To the extent that our colleges and universities can produce creative and broadly literate citizens empowered to navigate the vicissitudes of a knowledge economy driven by perpetual innovation as well as a commitment to undertake planetary stewardship, skepticism regarding the capacity of our democracy to successfully negotiate the perils of climate change may be unwarranted.

13 NEW AMERICA AND THE LANDSCAPE OF DEMOCRACY

Wellington Reiter

So, it doesn't take much to tip that over and get to the point where nothing can go downstream. And if you don't take it seriously now, if you think that you're going to avoid this . . . you're insane.
—Pat Mulroy, former general manager of the Southern Nevada Water Authority, speaking on the crisis of the Colorado River, CNN, August 17, 2022[1]

It is August 17, 2022, and the front page of the *New York Times* features yet another haunting photograph of the rapidly retreating water levels in Lake Powell and the parched terrain left behind. The headline references the much-anticipated intervention by the US Bureau of Reclamation to address the depletion of the Colorado River and the distinct possibility of "dead pool" situations in Lakes Powell and Mead, the two largest reservoirs in the United States.

Portrayals of a drought-stricken US West have become daily fixtures in our news feeds. Approximating lunar landscapes, albeit with previously submerged vessels jutting out of the drying mud, these images have become emblematic of human intervention into systems of scale and complexity that exceed our ability to manage them. The ramifications are certainly distressing for the forty million people who depend on the Colorado River for water, energy, agriculture, and recreation. But the scenes of depletion shake the confidence of the nation as a whole, prompting questions about the fragility of contemporary life and the assumptions on which it is built. Although it might appear otherwise, this particular situation was not unanticipated. In fact, it was foreshadowed more than 150 years ago, offering a case study of our resistance to difficult but critical information.

In this chapter I reference three journeys across this country—two that are historically relevant and a third that is just beginning. All speak to the value of direct observation of the environment we have created and how clues to the future surround us, if we dare to look. In combination with contemporary data analytics, *no society in history has ever enjoyed greater access to a predictable future*, especially as it regards the onset of climate change. The question is whether we can leverage this knowledge, come to terms with the unsettling implications, and make informed decisions for the benefit of generations to come.

IN SEARCH OF WATER IN THE ARID LANDS OF THE WEST

The story of John Wesley Powell's harrowing navigation of the Colorado River with wildly inadequate equipment in 1869 is well documented. Powell was an indefatigable citizen scientist focused on the land, the ecosystems it sponsors, and native American culture (he would eventually help to establish the United States Geological Survey and serve as the first director of the Bureau of Ethnology at the Smithsonian Institution). In the process, he came to understand the land west of the 100th meridian as exceptionally dry and a challenge for future development if not organized around the limited water resources available. His observations and concerns were captured in his comprehensive presentation to Congress, the *Report on the Lands of the Arid Region of the United States* (1878).[2]

Distilled from his reconnaissance throughout the West, the report may be best remembered for its remarkable map, one achieved without the benefit of the satellite imagery we take for granted today. Carefully drawn and color-coded, it offered a collection of interlocking jurisdictional units that corresponded to the natural topography and resulting watersheds. Rhetorically powerful, its conclusion was abundantly clear: as the country expands westward, be sure to organize human habitation in a format corresponding to the sources of water on which it will depend.

Historian John F. Ross characterized Powell's well-researched and authoritative position this way: "A future that did not contemplate reasonable (water) limits would be lured into swamps of unsustainability: shortages, endless litigation, the demands of infrastructure, feuding water politics—each one a threat to a democratic society the likes of which had never been attempted on so grand a scale."[3] Sensing a mismatch between supply and future demand, Powell tried to explain matters of hydrology to a Congress made up of representatives with minimal understanding of the West who were driven by constituent interests, election cycles, and profit motives rather than concerns for multigenerational benefits.

While awareness was raised, the motivation to act in accordance with the map was not. The nuanced cartography Powell provided stands in sharp contrast to the rudimentary shapes of the states we see today—what could fairly be described as a triumph of expediency over complexity. Reflecting nothing of the topography onto which they were drawn, these simple geometries demonstrate a lack of correspondence not only with the land but even with each other and the resources they share—watersheds being the most critical.

Some fifty years after Powell's first take on how water flowed through the West, seven states dependent on the Colorado River organized themselves into an upper and lower basin arrangement, with water allocations for each, codified in the Colorado River Compact (1922) and subsequent agreements known as the Law of the River. This reflected a desire for a regionally based governing body, not a federal agency, to capture the interests and responsibilities of those with a direct stake in the planning and outcomes. So far, so good.

However, the effectiveness of frameworks and maps are limited by the receptivity to the recommendations they offer, something Powell experienced firsthand. In their remarkable volume, *Science Be Dammed*, Eric Kuhn and John Fleck remind us that we often set aside hard-won knowledge and difficult decisions in favor of political accommodation. In the case of the Colorado Compact, an overestimation of what the river could actually deliver was built into the agreement's structure. As Kuhn and Fleck explain, "The decision-makers actually had available, had they chosen to use it, a relatively thorough, complete, and almost modern picture of the river's hydrology."

Regarding access to reliable estimates, the authors reiterate, "They had available, but chose not to use, data suggesting a much smaller river in the years prior to the eighteen years on which they were relying. Had they taken the science seriously, they almost certainly would have concluded that the Colorado River had less water than the common assumptions underpinning their race to develop the river."[4]

The result, a century later, are photographs of a diminished natural resource, a crisis in the reservoirs, and a predictable situation now exacerbated by climate change. To meet the moment, the Bureau of Reclamation is recommending massive user reductions, "between 2 and 4 million acre-feet of additional conservation . . . just to protect critical levels in 2023." It is likely to be an unachievable target, as it is equal to the total that some basin states receive on an annual basis. If these reductions are insisted upon, the impact on cities, industry, and especially agriculture will be profound. As recently reported in the *New York Times*, "The Colorado River is hurtling toward a social, political and environmental crisis at a pace that surpasses the Law of the River's ability to prevent it. In a world of less water, everyone who uses the river must adjust."[5]

While Powell could not possibly have covered every corner of the West on foot or horseback, in hindsight his overall advice was sound. Unsuccessful in his efforts to sway lawmakers, he nevertheless provided the country with a cautionary tale about how we gather, assess, disseminate, and act on critical data—or just as frequently fail to do so. Unfortunately, the crisis he had hoped to avoid is now building toward a painful climax.

HIGHWAYS FOR A NATION ON THE MOVE

The population of the United States after the Civil War was a little more than thirty million. Less than a century later, it had quintupled to 175 million across the lower forty-eight states. Powell's map offered well-reasoned advice about the distribution of new population centers and their associated demand for water. Of course, there were many other powerful drivers of urbanization (land value, resource discovery, industry, politics, and boosterism), which resulted in the scattering of cities across the continent. It soon became obvious that what they lacked, in addition to water, was connectivity.

Just as Powell's map and water advisories were a function of his travels, Eisenhower gained a vivid understanding of the country's transportation "system," such as it was, as part of the military's First Transcontinental Motor Convoy of 1919. This 3,200-mile, sixty-two-day, coast-to-coast excursion from the nation's capital to San Francisco was intended as an assessment of the defensive posture of the nation and the capacity of our infrastructure to support it. The test did not go well. More than three hundred men and eighty vehicles averaged less than six miles per hour and experienced some 230 "incidents" resulting in collapsed bridges, disabled equipment, and loss of manpower due to injury. What the trek revealed was the inadequacy, even impassability, of America's uncoordinated collection of roadways, all of which was detailed in his report. This stood in sharp contrast to Eisenhower's later assessment of German transportation networks, as outlined in his memoirs: "During World War II, I had seen the superlative system of German autobahn—[the] national highways crossing that country," an asset he would put to good use in the course of prosecuting the war effort.[6]

Given his direct observation of the advantages of high-quality roads, it is not surprising that then-general and later President Eisenhower would take a strong interest in equipping the United States with a robust transportation network. The result was the Federal-Aid Highway Act of 1956, resulting in the superimposition of a 41,000-mile ribbon of concrete across the continental United States. The objectives were clear: support national defense, expand commerce, and facilitate the western migration that Powell had anticipated nearly a century earlier. The system, which we depend on today, is a distinctly twentieth-century creation and reflects a country leveraging capabilities honed in the course of global conflict and confident in its position on the world stage (the benefits of having the Supreme Allied Commander as president is apparent).

Not simply a means to an end, highways were the embodiment of a nation in a hurry, a capitalist society in full bloom, and remain, for better and for worse, a testament to who we are and what we value. While this was a grand gesture at the federal level, it also served to elevate the importance of cities, especially in terms of their contribution to an industrial economy and nonagricultural employment. The very idea of the suburbs was a by-product of these new roads. In many ways, the interstate system, along with the automobile, production housing, and air conditioning, generated the script for the American dream.

On the other hand, it would be fair to ask whose dream this was, and how carefully the possible outcomes were researched. Threading multilane roads through the fabric of already established cities would inevitably be disruptive and would irreparably change the character of once-walkable neighborhoods. It would also be discriminatory. The eminent urbanist Lewis Mumford, writing in *The Highway and the City*, was certain of the fallout:

> When the American people, through their Congress, voted . . . for a twenty-six-billion-dollar highway program, the most charitable thing to assume about this action is that they hadn't the faintest notion of what they were doing. Within the next fifteen years they will doubtless find out; but by that time it will be too late to correct all the damage to our cities and our countryside.

Mumford, clearly speaking with passion and urgency, went on:

> Yet if someone had foretold these consequences, . . . it is doubtful whether our countrymen would have listened long enough to understand; or would even have been able to change their minds if they did understand.[7]

Of course, we did change our minds, but only after the realities of this grand experiment became evident, especially in our cities. As a result, we have been removing highways from urban centers (think of the Big Dig in Boston and the Embarcadero in San Francisco). As Mumford forecast, while affording freedom of movement for the nation as a whole, these immense constructions diminished the quality of life in local communities, disregarded the voices of those who lived there and, in the process, rendered some racial inequities in built form. Less noticed until recently was how highways locked in our addiction to fossil fuels and the thousands of metric tons of CO_2 pushed into the atmosphere annually—a major contributor to climate change.

Nevertheless, the 1956 Highway Act—in spite of its deficiencies—is frequently invoked as a positive display of government executing on ambitious goals, and what is possible when consensus is an option (the bill passed the House of Representatives by a vote of 388 to 19, a ratio beyond imagination in our current era of party-over-country politics). Generations later, the Infrastructure Investment and Jobs Act, of 2021, is dedicating federal resources to remedy the oversights and blatant mistakes of past projects with the hope of bringing inclusion, well-being, and environmental issues to the front of the planning process. We can learn.

TEN ACROSS AND THE KNOWABLE FUTURE

Access to water resources in the arid West and the expeditious movement of goods and people across the continent remain active concerns as we reach thresholds of viability under existing infrastructure models. Presented with the vast US landscape and a population beginning to fill it out, Powell and Eisenhower responded in keeping with the paradigms of their respective eras, senses of duty, and political circumstances. Ignoring the artificiality of state boundaries, they were thinking at the scale of the nation and for the long term. Most importantly, they imagined alternative futures that were informed by direct observation of the country in the course of their respective fact-finding expeditions of 1869 and 1919.

Now it is 2023, and we need to think big again and to revisit the land that we have bent to our will and altered to an extraordinary degree. Specifically, we need a clear-eyed examination of all aspects of the built environment in order to pressure test them against the unavoidable forces of climate change. The authors of the recent study *Megaregions and America's Future* offer a concise description of the challenge, one that unites issues at the intersection of place and governance:

> The United States has daunting economic, demographic, climatic, environmental, and spatial development challenges. Many of these ignore existing jurisdictional boundaries and spread over large regions. The recent near collapse of the Texas power grid because of a climate-induced cold snap and poor planning, for example, underscores the urgent need to design and manage this and other infrastructure systems at a scale beyond even that of our largest states. The complexities of coming decades are likely to evolve in dynamic special processes that our current fragmented political geographies and rigid institutional systems will not be able to adequately address.[8]

It is time for another urgent fact-finding expedition in search of something less tangible than sustainable water resources or vehicular mobility—although we need revised versions of both. Instead, the quest is for confirmation of our capacity to make informed, brave, fact-based decisions in response to a rapidly changing world along with the societal will to back them up. This will be a test not only of our inventiveness but of our democracy, as well.

The Ten Across initiative is an inquiry into our ability to gather well-researched data, understand and embrace the implications, and alter our present trajectory or, at minimum, stave off catastrophe.[9] Roughly tracking US Interstate 10 across 2,400-miles, from Los Angeles, California, to Jacksonville, Florida, one can observe through this initiative the full spectrum of symptoms affiliated with a warming planet in their extremes: increased triple-digit temperatures, unrelenting drought, year-round fire "seasons," regularly occurring "five-hundred-year" storms, devastating floods, and thousands of square miles of land loss due to sea level rise. In epidemiological terms, this would be described as a "hot zone," a site where conditions are morphing at a pace beyond our ability to manage the disruptions. Counterintuitively, this corridor also features the fastest-growing and most diverse cities and counties in the nation.

The Ten Across transect is already the proving ground for the most critical issues of our time. With the "energy capital of the world" at its center (Houston), this is where the transition from fossil to renewable sources must take place. Also being tested is the persistence of mechanically enabled cities, whether maintained below sea level (New Orleans) or as islands in the scorching desert (Phoenix and Tucson). Globalization is another ongoing experiment, the pros and cons of which are exposed via the massive ports in the geography (Los Angeles and Baton Rouge) and, of course, by migration along the much-disputed border (El Paso/Ciudad Juarez). All of these conditions are either contributors to or are being significantly altered by climate change. The world is watching to see whether a "first-world" democracy can effectively grapple with them. If not, the implications are dire.

Of course, the political context is everything as it regards our adaptability to climate change, including receptivity to difficult information (Powell) or achievement of consensus around a plan of action (Eisenhower). Coincident with the issues listed above, the Ten Across geography will be the most active stage for debating the future of democracy in the United States, the outcome of which can no longer be assumed. Included in our study area are the three largest states (California, Texas, and Florida), which total 27 percent of the House of Representatives and exert an outsized influence on all aspects of the national dialogue—especially the environment. With some legislatures along the Ten Across corridor questioning the reality of climate change while also casting doubt on the validity of representative government, there is reason to doubt our capacity to respond proactively to meaningful data and/or achieve consensus around extremely difficult decisions. Nevertheless, given the urgency of the trendlines and the irrefutable evidence, it is likely that climate and democracy will be fully intertwined by the time of the 2024 election, one that political observers suggest presents as an existential test for governance in the United States.

In all of its complexities and contradictions, the Ten Across transect is where the future has come early. It is a uniquely American landscape, one with abrupt juxtapositions: unparalleled beauty threatened by industrial degradation; rapid urbanization competing with essential agriculture; profound new wealth hovering over long-standing inequity; the California model versus red-state conservatism. It is, for better or worse, the New America, and we would do well to understand it thoroughly.

If he were alive today, Powell—the explorer, geographer, and anthropologist—would be traveling repeatedly over this transect, equipped with the latest instrumentation and data-crunching analytics, searching for evidence of our capacity to respond in an informed manner to the adaptation that will be required and to generate new narratives based on a knowable future. This is, in essence, what the Ten Across project will be pursuing in earnest.

In his prescient book *Time to Start Thinking: America in the Age of Descent* (2012), journalist Edward Luce suggests that the United States cannot continue to ride on the coattails of the "greatest generation" (Eisenhower's) and take its position in the world for granted or contend with the unprecedented challenges before us in our present format. Written a decade ago, this message rings even more true today. Luce observes, "When a country's narratives become captivated by the past, they rob the present of the scrutiny it deserves. They also tend to shortchange the future."[10]

The world we will soon inhabit will severely test our expectations, lifestyles, systems of governance, and societal cohesion. Having contributed significantly to the problem, our goal must be to slow further degradation of our environment and imagine new ways of coping in an effective, inclusive, and innovative manner. Thanks to contemporary computational capacity, Earth-observing technologies, artificial intelligence, and ubiquitous, first-person reporting from anywhere on the globe, *no society has ever enjoyed greater access to the future*. We have no excuses. The critical trendlines associated with climate change are bountiful, well-sourced, documented, and distributed; the question is whether we can handle the implications and convert them in into the requisite sense of urgency.

Today, Lake Mead, the reservoir created by the Hoover Dam is at approximately 25 percent of its intended capacity, the lowest point since it was filled in 1934. Recent meetings of the officials charged with managing this vital, shared resource have discussed measures to avert the potential catastrophe of a "dead pool" condition. Maintaining the flow of the Colorado River presents a great technical challenge—yet at its core, it is a matter of human nature and will. Sculptor Oskar J. W. Hansen, creator of the winged figures atop the Hoover Dam, called it "a monument to collective genius exerting itself in community efforts around a common need or ideal."[11] Moving forward, we need to be precise in the measurement of our available resources and responsive to the implications of a changing climate. Yet perhaps most importantly, we will again need to draw on our "collective genius" to meet our responsibility to future generations.

14 WATERING THE ROOTS OF DEMOCRACY

Richard Louv

At first glance, the topics of nature, children, and democracy may not seem related. But they are. To explain this relationship, let me start with a story I've told often in recent years.

More than two decades ago, I visited Southwood Elementary, the grade school I attended when I was a boy growing up in Raytown, Missouri. From the windows of the classroom, I could see the same trees, branches bare, that I had watched dip and sway when I was a boy.

In that classroom, I asked fifth graders about their relationship with nature. Many of them offered the now-typical response. On those occasions when they were outside, the students were more likely to be playing soccer or some other adult-organized sport. More than a few said they preferred playing indoors, because, as a third grader in another school once told me, "That's where all the electrical outlets are."

Then an eleven-year-old stood up. She was wearing a plain print dress and an intensely serious expression. Earlier, a teacher had described her as "our little poet."

"When I'm in the woods," the little poet said, "I feel like I'm in my mother's shoes."

To her, nature represented beauty, refuge, and healing. "It's so peaceful out there and the air smells so good. For me, it's completely different there. It's your own time. Sometimes I go there when I'm mad—and then, just with the peacefulness, I'm better. I can come back home happy, and my mom doesn't even know why."

She paused, then said, "I had a place. There was a big waterfall and a creek on one side of it. I'd dug a big hole there, and sometimes I'd take a tent back there, or a blanket, and just lay down in the hole, and look up at the trees and sky.

Sometimes I'd fall asleep back in there. I just felt free; it was like my place, and I could do what I wanted, with nobody to stop me. I used to go down there almost every day."

The young poet's face flushed. Her voice thickened. "And then they just cut the woods down. It was like they cut down part of me."

If the eminent biologist E. O. Wilson's "biophilia hypothesis" is correct—that the human attraction to the rest of nature is genetically hard-wired within us—then our young poet's heartfelt statement was more than metaphor. When she referred to her woods as "part of me," she was describing her primal biology, her sense of wonder, her sense of belonging to a larger community of nature, including human beings.

Not long ago, I imagined the poet's path to her special place among the trees.

Perhaps she encountered a kind neighbor standing in his garage who noticed one day that she was crying, and spoke to her softly, and loaned her a book. She might have studied the houses, the corner store, the little park. Maybe she saw a father standing in his driveway watching the gathering clouds. And then the houses fell behind, and she walked through a cornfield into the trees, where pocket mice and mourning doves and a garter snake in a shadow communicated in their ways. Both the human neighborhood and the neighborhood of animals were part of her community. She knew them both.

In recent decades, such intimate knowledge of community has faded. Too many Americans have withdrawn from nature and each other. School districts have cut back on or eliminated recess. Forests have been degraded. Garage doors are closed, strangers feared, the woods avoided. This unprecedented social experiment, this mass migration from the real to the virtual, has taken a toll on our health, on nature, and on democracy.

To develop a sense of the larger community, a child or an adult must step outside, must get to know the neighbors, both human and other-than-human.

As the poet Wendell Berry has said, "If you don't know where you are you don't know who you are."

When we connect children to nature, we water the roots of democracy.

HOW NATURE BUILDS VALUES AND CONDITIONS REQUIRED IN A DEMOCRACY

The democratic process is not only about voting, political consultants, polls, winners and losers. *Social* democracy is about social capital; how we treat each other, whether we view others as equals. And democracy is about health, of people and of the rest of nature.

Equitable Access to Nature Builds a Healthier Polity

During the past fifteen years, a growing body of research has linked our connection to nature to reductions in vitamin D deficiency, myopia, obesity, diastolic blood pressure, stress-related salivary cortisol, heart rate, diabetes, and mood disorders.[1] One study reported that for young adults, the more nature they experienced over a fourteen-day period, the more life satisfaction they felt daily. In short, time in nature heals. Nature exposure in childhood or adolescence might also protect against cognitive decline and mental health issues later in life.

Access to nature can literally be a matter of life and death. In 2017, a study published in the prestigious medical journal *The Lancet Planetary Health* suggested that people who live in green neighborhoods live longer than those with little nature nearby.

And in 2018, UK researchers reviewed studies involving more than 290 million people of all ages from twenty countries. Their analysis confirmed what other researchers had been reporting. They also found that green space exposure reduces the risk of preterm birth, premature death, and high blood pressure—all of which disproportionately affect people of color.

A Greener Society Contributes to a Better Educated Voting Public

In education, time spent in natural settings, including outdoor classrooms, can improve cognitive functioning, reduce attention-deficit disorder, and raise test scores; it helps teachers avoid early burnout and builds resiliency in children and their communities.

Cognitive and behavioral benefits accrue well beyond school boundaries. In inner-city housing projects in Chicago, investigators found that the presence of trees outside apartment buildings were predictors of certain behaviors: less procrastination, better coping skills, greater self-discipline among girls, better social relationships, and less violence.

Learning outside suits a wider array of students. One study found that so-called at-risk students in weeklong outdoor camp settings scored significantly better on science testing than in the typical classroom.

In 2019, the noted University of Illinois professor Ming Kuo and her colleagues published a systematic review of nature-related education research in the journal *Frontiers in Psychology.* They conclude that greener schools—ones that, for example, offer a natural space for play and learning, take students on field trips to natural areas, and bring nature into the classroom—reduce stress, boost cognitive functioning, and may raise standardized test scores and graduation rates. According to Kuo, "Report after report—from independent observers as well as participants themselves—indicate shifts in perseverance, problem solving, critical thinking, leadership, teamwork and resilience."[2]

All of these qualities are essential to the building of democratic values.

Immersion in Nature Nurtures Empathy and Builds Social Capital

No democracy can long endure if its citizens are unable to tolerate different points of view, or to walk in another voter's shoes. The political and cultural polarization of the United States can be blamed on universities, Twitter and Facebook, FOX News, or the language police. But, for whatever reason, our society suffers from a severe empathy shortage.

How do we nurture empathy? Numerous studies have documented how relationships with pets and wildlife builds empathy and compassion in children. Research suggests that nearby nature can reduce neighborhood and domestic violence. Greener communities and learning environments, including the presence of animals in schools, also builds empathy and social capital—the glue that holds a society together.

For example, children in a natural play space are more likely to invite children of other races or genders to play with them, to be inclusive and fair, as compared to a typical asphalt playground. They're also more likely to invent their own games, which builds executive function: the ability to make one's own decisions, to be entrepreneurial—still another economic benefit.

Biophilic Design and Nearby Nature Strengthens the Economy

Energy efficiency is not the only benefit of green buildings. Designers and architects in the growing field of biophilic architects are weaving plants, trees, and other natural elements in and around workplaces and schools. The walls become vertical gardens, courtyards become botanical gardens, surrounding land becomes forest. Studies of biophilic buildings have shown that workers there are more productive and creative; sick time and absenteeism are reduced.

In the 1970s, schools were being built with windowless classrooms (the theory being that a view of the outdoors was distracting). Multiple studies have shown the importance of natural light to learning. One 2001 study of elementary schools revealed that students in classrooms with the most diffuse daylight showed a 21 percent improvement in learning rates compared to students in classrooms with the least daylight.[3]

If whole cities were to be *equitably* designed or retrofitted through biophilic principles, the psychological, educational, and economic benefits would boost the economy and democracy.

Still, our connection to the natural world is threatened as never before, and that threat undermines the gifts of nature and democracy.

Nature Connection Reduces Environmental Despair and Increases Action

In recent years, health-care researchers have warned about the growing "epidemic of human loneliness," as some describe it. This epidemic—or pandemic—is older than the COVID-19 pandemic, runs parallel to it, and is exacerbated by it. As a contributor to major illnesses and death, social isolation now ranks with obesity and smoking. Disturbingly, one study found that the younger the generation, the lonelier its members are. That's a dramatic reversal from prior decades. What does it say about a society in which the younger a person is, the lonelier they feel?

Poor urban design that discourages walking and preservation of nearby nature, an amplified fear of stranger danger, not knowing our neighbors or our neighborhoods, and antisocial media are some of the reasons most cited for the loneliness epidemic.

Not coincidentally, they happen to be some of the same factors that have divided children from nature—and have separated Americans into insular political tribes.

In *Our Wild Calling*, I make the case that this aloneness is rooted in an older and even deeper isolation: *species* loneliness. As human beings, we are desperate to feel that we are not alone in the universe. This yearning has religious implications, but it also suggests that the path back to each other leads through woods and fields, through the greater community.

We're not alone, if we pay attention. Early in the pandemic, people sequestered in their homes and apartments found solace and took comfort from rescued pets and from the presence of birds and other wild animals just outside their windows. In the company of the dog rescued from a shelter, the acorn woodpeckers tending their nest, the fox crossing the yard and making eye contact, people felt less alone. As the restrictions lifted, adults (but surprisingly few children) instinctively rushed to parks and trails for their health and their sanity.

To some, this recognition felt like coming home.

In 2019, a few months before the pandemic officially began, raging firestorms raced across the Australian continent, destroying more than 20 percent of that continent's forests. The images from that cataclysm were both painful and inspiring.

From afar, we watched people on bikes charge through scorched and burning forests to rescue suffering animals. We saw koalas climb onto human laps to reach the bottles. We witnessed the courage of the "carers," as the NGO Animals Australia called the compassionate people who had "lost everything" but headed "out with only the clothes on their backs to help injured and burned animals." The images were at once painful and uplifting. We marveled at these demonstrations of human kindness.

In time, the transitory rains arrived, the fires receded as a virus spread, and those powerful images inevitably began to fade from public consciousness. Environmental conditions continued to decline.

What will it take to move our species to act on climate disruption, biodiversity collapse, unending pandemics, and human isolation? A reliance on facts and logic is clearly not enough. We need at least two additional ways to make the case. The first is love—deep emotional attachment to the nature around us. The second element is imaginative hope. In his recent book, *Earth Emotions*, Australian eco-philosopher Glenn Albrecht argues that only "a shift in the baseline of emotions and values has worked" to transform facts into action in other areas, such as feminism, same-sex marriage, and, to a degree, racial inequities. The reason these good causes have made at least some progress in Albrecht's Australia and other countries is because "they revolved around the issue of love." This is why the images of suffering animals and human heroism in Australia are so important. They remind us, at least for a while, that we belong to a larger family, one worth loving.

In a 1787 letter to the son of John Adams, Thomas Jefferson issued this now-famous quote: "The tree of liberty must be refreshed from time to time with the blood of patriots and tyrants." Jefferson added, "Our [Constitutional] Convention has been too much impressed by the insurrection of Massachusetts [Shay's Rebellion, against taxes] and . . . they are setting up a kite [a raptor] to keep the hen yard in order."

It's been a while since we've heard a US politician utter any statement packed with so many allusions to the natural world. Doing so was common in Jefferson's time. His quote is misused today, to rationalize violence. Given the climate emergency and the fires and droughts that it brings, today, the tree of liberty requires more water. And care. And around it, a new forest. Conservation is not enough. Now we need to create nature. And then love it.

IMAGINE A NEWER WORLD

Not surprisingly, many Americans, young and old, find it increasingly difficult to imagine a brighter future. What we need now, most of all, is imaginative hope—the ability to foresee and then create a future that is not only sustainable, but nature-rich, beautiful, healthy for the children of all species. Though the moment may pass, young people (and older ones too) are primed to do just that. Despite the apparent odds, despite the irony that so many members of this young generation have been denied the personal connection experienced by prior generations.

Belief in democracy requires a reasonable faith that it works. That faith now wavers, as government seems inadequate to confront the new Four Horsemen of the Apocalypse: climate change; biodiversity collapse; zoonotic pandemics (which can be buffered by biodiversity); and human loneliness. These forces are interdependent; they ride one horse. We cannot do much about any one of them without confronting all of them.

One sign of hope is the emerging new nature movement, which is dedicated to closing the gap between children and nature. And it's more than that.

Reconnecting children and nature may be the last cause in America that transcends political, religious, racial, and professional barriers; it brings people to the same table who usually do not want to be in the same room. Again and again, I have seen conservatives and liberals, physicians and educators, conservationists and developers, and many others work together for this cause. *No one* wants to be in the last generation where it's considered normal for a child to lay under a tree in the woods and watch clouds become the faces of the future.

We now see pediatricians in the United States and Canada, as well as other countries, writing prescriptions for nature time. We see a growing green schoolyard movement and the development of nature-smart libraries, as well as the impressive growth of nature-based schools and outdoor classrooms. We see architects using biophilic design for workplaces and schools and, potentially, whole cities. We see mayors and other municipal leaders striving to make their cities nature-rich for the children of all races and backgrounds.

We see other nations, including China, Brazil, Australia, and Brazil, creating national campaigns to connect children to nature. We see parents, grandparents, and people without children committing themselves to a shared future in the natural community. They know intuitively that their children and grandchildren are inseparable from the fate of the natural world. The new nature movement is about imagining a newer world.

A healthy democracy requires a public freed from the current dystopian trance, a people capable of imagining a better future—not just a survivable future.

It's still possible to imagine that newer world. The multiple simultaneous solutions required to simultaneously address environmental collapse and human loneliness can also help reduce racism, violence, poverty, and improve human health. These challenges are intertwined and will not be met without widespread action by individuals and large-scale political action.

Success will depend on a rising constituency and a more muscular environmental movement—hope with teeth—and people, especially young people, capable of envisioning new images of a beautiful, nature-filled future, where agriculture regenerates nature; where biophilic architecture and design create workplaces, schools, and homes that not only conserve energy but produce human energy; where cities become engines of biodiversity and human health; where vast forests are planted to absorb carbon and support human and more-than-human health; where the definition of green jobs expands beyond those associated with energy efficiency to ones that connect the human species to the rest of nature; places of worship call us to care for the Creation, where reciprocity and empathy survive.

At its core, this new nature movement and the impulses that drive it are profoundly democratic. The eco-theologian Thomas Berry once wrote, "A degraded habitat will produce degraded humans. If there is to be any true progress, then the entire life community must progress." At some level, our founding fathers and mothers understood that, given their belief in inalienable and natural rights.

For more than a decade, some of us have argued that a positive connection to nature should be considered a human right—because a positive connection to nature is so fundamental to our health, to our humanity, to democracy, and to the health of the Earth itself.

In 2009, in *Orion* magazine, I wrote,

> This truth must become evident: *We can truly care for nature and ourselves only if we prize the larger community; only if we see ourselves and nature as inseparable, only if we love ourselves as part of nature, only if we believe that human beings*

have a right to the gifts of nature, undestroyed. The young poet in Raytown may not have had a specific right to a particular tree, but she and all people have the inalienable right to be with other life; to liberty, which cannot be realized under protective house arrest; and to the pursuit of happiness, which is made whole by the natural world.

We're seeing progress toward achieving this goal. In 2012, the International Union for Conservation of Nature, the world's largest network of conservation NGOs, passed a resolution declaring nature connection a human right for all children. Eventually, this right must be recognized by every culture, along with the rights of nature itself.

The community of life, to which we belong, will depend on that recognition.

AFTERWORD

Kim Stanley Robinson

This collection of essays performs an act of cognitive mapping. As used by Fredric Jameson, cognitive mapping is a name for a kind of ideology formation, using the word ideology in its Althusserian meaning—in other words, not the false consciousness that other people have, but rather one's imaginative relationship to the real conditions of existence, a necessarily creative and ongoing construction that allows us to make sense of the "blooming buzzing confusion" of our sensory inputs, as well as the modern world's overwhelming onslaught of information. In this interpretation, having an ideology is crucial for deciding what to do.

Maps helps to situate us in space; cognitive maps presumably exist in our minds. I think the adjective is an attempt to suggest that cognitive maps are active and thus have a temporal element. They are about not just space, but time. We try to locate ourselves in the flow of history.

These essays attempt to map our moment in useful ways. But maps are never the territory. Some parts of the territory get mapped, others not; maps have not just their terrae incognitae, but their principles of selection. These become algorithms that then get applied. Possibly ideology is a kind of algorithm, in which case it's always helpful to wonder what's been left out of the calculation.

This book asserts that democracy is under threat, and that climate change is endangering the biosphere that sustains all living creatures. Both cases are easy to make on their own. Many of the essays further suggest that these two situations are linked or intertwined, because democratic governments do a better job of protecting the environment than authoritarian governments. Evidence for this is mixed, but most of the historical evidence does seem to support the idea that our best chance of dealing successfully with the oncoming climate catastrophe is by democratic means.

But why should this be so? And aren't there some arguments out there saying that we need to deal with climate change so desperately that we can't worry about whether we can win a democratic argument about it or not? There is mention in this book of "climate Leviathan," which seems to be naming an intensification of capitalist power that somehow cures climate change while keeping the current economic system, thus increasing inequality among humans. I've also heard the term "eco-fascism," which seems to refer to the state taking over the economy and ordering effective climate actions to happen. Of course the state and finance work hand in hand (and are maybe conjoined twins), but they are always arm-wrestling for ultimate control of society. "Climate Leviathan" seems to emphasize the market, though Leviathan traditionally refers to the state, while "eco-fascism" seems to emphasize the state, though fascism was also the name for a particular kind of economy. So these terms are slippery, perhaps because they both embody analogies to earlier historical situations that are not actually very analogical to our current moment, the climate crisis being unprecedented: we have never before begun a mass extinction event that will wreck our civilization.

If we do choose to think by way of historical analogy, momentarily, I think we need to recall World War II. In that existential struggle, the democracies among the Allied countries elected governments that took on authoritarian powers—temporarily in some cases, permanently in others. So we now say the democracies won the war; but those governments took over the economies of their countries in order to do that, and there were sizeable minorities (or even slight majorities) in those countries who were so alienated by this takeover that soon after the war ended, opposition parties were voted in and dismantled some of the authoritarian powers that government had taken on in the war—and even some powers that dated from the New Deal. Even so, do we perhaps need that kind of government takeover of private capital and private social life now, facing a mass extinction event that will hammer civilization in ways that could be even more damaging than World War II?

This needs to be discussed.

In any case, if democracy is the right political system to deal with climate change, it needs to be functioning well. Or we could even say that it has to be made real at last, or re-created for our time, by a continuous intense effort. Otherwise it is just a name for a false front covering the real government, which could be described as a capitalist oligarchy that has bought the ostensibly democratic mechanisms of power and suborned them to the oligarchy's purposes. These purposes seem careless of climate change's biospheric dangers, even stupidly unaware of these dangers, which threaten all living creatures, oligarchs included. Civilization could fall apart under the pressures and blows of climate change, and the fantasy of escape to a safe place outside civilization, to New Zealand or Mars or just the local guarded mansion, is a truly ridiculous fantasy, ideology in the older definition of false consciousness; delusional; and yet it must have an appeal for some of the people rich enough to entertain it. Meanwhile, the rest of us need to deal. And that will take democracy.

Or rather: *good governance*. Good governance is somewhat rare in history, and it always depends crucially on the consent of the governed, so maybe this is one definition of democracy, perhaps slightly tautological, but not exactly the same as the right to vote. I notice that some people speak of "good political representation" rather than democracy, as a name for government working well for the people being governed. This shift in terminology is possibly an attempt to deal with the example of China, whose citizens often report more satisfaction with their political representation than the citizens of Western democracies. How do we account for that? Indoctrination, bad polling, weird framing? Or simply the feeling people have about their political representation, whether or not they get to vote for it in elections? This isn't clear, and it's an issue that is often discussed, simply as part of what political science does a discipline. It's an issue worth pondering as you come out of this book and think about what it has said.

Another issue: I think the discussion here of "technocracy" is insufficient, especially if it means any kind of put-down of expertise, experts, and science as a part of democratic governance. We need science, and we need our experts and technocrats. If the general public approved of a "scientific meritocracy" (as H. G. Wells called it) running human affairs, then this would perhaps be a version of democracy appropriate to our current crisis, or any time at all.

And it may be that "the climate crisis" is not a good name either, given the entanglement of catastrophes now well begun. *The polycrisis*, a term popularized recently by Adam Tooze, is a name that attempts to include more problems than just climate change, such as extinctions, biosphere collapse, new pandemic diseases, shortages of food, fresh water, and various minerals; also the inadequacy of the international nation-state order and the global economic system to cope with all this, leading to mass refugee flows, social unrest, and possible societal breakdown. All these problems tangle together in a way that justifies the new name polycrisis. So to be complete, a book that cognitively mapped our situation would need to include all these things.

But that would be a very long book, made quickly obsolete by the tumultuous onrush of new events and information. Here the editors have decided to confine the essay topics to two big parts of the polycrisis, and all the essays are admirably up to date (which is to say, to late 2022). More than that would be too much to ask of the editors and their writers, but perhaps not too much to ask of MIT Press, this being the kind of ongoing project they are good at. Sequels, please!

Back to democracy. What do we mean by it? This is becoming a crucial question in history (again). Once you get above the scale of the town hall meeting, in which every citizen votes on every measure (in theory), there is always representation and administration. So if people generally approve of the system by which their representatives are chosen, is that democracy, or is it (just) hegemony?

There's a mention in this book of government by people chosen by lottery from the population (sortition); a good idea, which I've portrayed once or twice in novels. This would seem to be a clear case of Lincoln's "government of the people, by the people, and for the people." Recently I've seen Lincoln's description or slogan or prescription usefully collapsed to the "of by for" movement. This is an encouraging sign of pushback against the version of democracy wrested away from the people and now calcified in US law and practice. I've always maintained that the final words of Lincoln's Gettysburg

Address constitute a utopian short story, compressed to eighteen words: "that government of the people, by the people, and for the people, shall not perish from this Earth." It's the word *shall*, its imperative—this is what makes it a utopian call to action. We are exhorted to make this situation real, in a perpetual and ongoing act of creation. And this exhortation was made in the midst of a great civil war that was not at that point so obviously about the fate of democracy per se. Lincoln in that fraught and tragic moment wrote a great and enduring utopian story, shifting the grounds of that particular war, and thus also of American history, which in its origins was not by any means a completely democratic project.

The chapter by Kuh and May in this collection, about our constitution, is good on this question of whether that ultimate law is up to the challenges of our polycrisis—at first look, maybe not; but Kuh and May point out that certain clauses in the constitution can be interpreted to give our tripartite government the power to deal as necessary. This realization, combined with Becker's admirable to-do list of political reforms, constitute together a kind of campaign platform, which the Democratic Party is the obvious candidate to declare for and to enact. It's coherent, principled, wide-ranging, and within the reach of ordinary politics in the United States today—if only just. But a working majority is always within reach, and it's called a working majority because it works. It could be done.

What would help it to happen? An equally wide-ranging and coherent economic program. Always in these discussions we are (or should be) talking about political economy, not politics pure and simple; politics is only one part of a larger process, and without the economic element included, there is no real impetus for change. In short, justice needs to be on the table; and that's an economic matter. Democracy that allows for vast economic inequality is a bad joke, a false front. We need to take the title of one of Thomas Piketty's recent books seriously: *Time for Socialism*. I say this: real democracy means public ownership and administration of all the necessities, especially energy, housing, food, education, employment, health care, and money. Add sharp progressive taxation on both income and assets to this kind of nationalization, and that would be democracy with some teeth to it, and some real appeal for citizens. Call it social democracy, or democratic socialism, or socialism with American characteristics, or public utility districts, or justice, or whatever. The economic element must be included for democracy to be real. One of these essays mentioned Robert Dahl; his *A Preface to Economic Democracy* is one text outlining this project. There are many others.

So, as Becker suggests at the start of his great to-do list: "nationalize fossil fuels." Then on from there. Government is in effect the biggest company of all, and it belongs to the people, who also make the laws that regulate and govern all smaller companies. If government was also "the employer of last resort," offering a job to anyone who wanted one, at a salary adequate to a good life, then private companies would have to match that adequacy to get any workers; this would establish a universal floor. Then progressive taxation would remove the possibility of obscenely gargantuan private individual wealth, creating thereby a wage ratio perhaps comparable to that of the US Navy, which is to say, one to eight; but let's call it one to ten, for simplicity's sake, for ease of calculation, and so on. A floor of comfortable secure adequacy, a ceiling of ten times that adequacy (which is already immense luxury): this would be a political economy worth articulating, arguing for, believing in, fighting for—and legislating, with the support of a majority that might grow quite large when the program was understood. Democracy in action, for economic justice first. After that, this same political economy would come to understand that the health of the biosphere is synonymous with the health of the body politic: not just a good idea, but necessary. The two parts of the program would gain power by being paired.

What else would help forward this program? Chapter 12 by Crow and Dabars on the role of the universities in spreading awareness, education, and capability to act with expertise fills a crucial gap in the story of how we can get there. We need to tell the story clearly and spread the word as far and wide as we can, and this is education in its best sense.

This collection provides an education and a program to be enacted. Thanks to the writers and editors for their clarity and their good work. Go little book!

Acknowledgments

This book is the result of collaboration among dozens of people across many organizations—all of them passionate about preserving and improving democracy and undaunted by the challenges posed by a changing climate. I am grateful to all for their diligence, commitment, and friendship. In the Colorado-based Climate/Democracy Initiative, my thanks to John Powers, Spike Buckley, Peter Brown, and Michael Giles as well as the members of our advisory group: Fritz Mayer, Bill Ritter, Brenna Simmons-St. Ong, George Sparks, Paul Teske, and Ben Withers, and Edward Barbier.

At Arizona State University, President Michael Crow has provided inestimable help and inspiration. His leadership in redefining higher education relative to the issues of climate, justice, expanding opportunity, and systemic innovation is extraordinary. Thanks also to the members of the ASU climate and democracy working group: Dean Patrick Kenney, Richard Amesbury, Carol Andrade, Battinto Batts, Bill Brandt, Ann Florini, Magda Hinojosa, Duke Reiter, Peter Schlosser, and Iveta Silova.

We could not have had a better or more supportive editor than Beth Clevenger at the MIT Press. From the start she encouraged, critiqued, and coached, all with professionalism, humor, and kindness.

This book has a history that began in November 2017 with a three-day Conference on the State of American Democracy at Oberlin College. Thirty-two speakers and panelists including Jane Mayer, David Daley, Jessica Tuchman Mathews, Jonathan Alter, Bill Ritter, Tim Egan, Gus Speth, Akhil Reed Amar, and the Reverend William Barber deliberated on our plight. Our goal was to clarify the origins of the election of 2016 and chart a path forward to repair and strengthen democratic institutions. In *Democracy Unchained* (The New Press, 2020), thirty-nine leading scholars, public intellectuals, and political leaders, including Bill McKibben, Ganesh Sitaraman, Ras Baraka, Jacob Hacker, and Paul Pierson explored how to repair and strengthen democratic institutions. My coeditors Bill Becker, Bakari Kitwana, and Andrew Gumbel were masterful editors and great colleagues as well.

The COVID pandemic forced us to cancel plans for fourteen events across the country with various teams of the contributors. Dan Moulthrop, president of the Cleveland City Club, and Graham Veysey of NorthWater Productions stepped in to help create eleven online events beginning at the National Cathedral in September 2020. Those conversations with ninety-eight participants including Danielle Allen, Ian Bassin, Preet Bharara, David Brooks, Sherrod Brown, James Clyburn, Russ Feingold, Catherine Flowers, Stephen Heintz, Maria Hinojosa, Bruce Jennings, Van Jones, Juliette Kayyem, Jill Lepore, Richard Louv, Nancy MacLean, Yascha Mounk, Trevor Potter, Kevin Roose, Tim Snyder, Max Stier, Pete Wehner, Tim Wirth, Terry Tempest Williams, Sally Yates, and Shoshana Zuboff, and reached an audience estimated at one million viewers.

None of this would have been possible without the good work of two friends and colleagues—neighbors, in fact: Jane Mathison and Julie Min. I have yet to meet two more competent, thoughtful, and perceptive people, and they do it all with a smile. Many others helped along the way, including Marvin Krislov, Rev. Doctor Andrew Barnett, Mary Evelyn Tucker, John Grim, the Right Reverend Randy Hollerith, and Michelle Dibblee. For their generous support I thank Adam and Melony Lewis and Andrew McIlwaine and Grant Oliphant at the Heinz Endowments. Above all I owe an unpayable debt to Elaine, Michael, Daniel, Ann, Lewis, Molly, Noreen, Ruby Kate, and Maggie who are a gyroscope of love, laughter, and hope in this troubled world.

Notes

INTRODUCTION

1. Kisei Tanaka and Kyle Van Houtan, “The Recent Normalization of Historical Marine Heat Extremes,” *PLOS Climate*, February 1, 2022; Alastair Graham et al., “Rapid Retreat of Thwaites Glacier,” *Nature Geoscience* 5 (September 2022).

2. That number does not include the effects of other heat-trapping gases measured in CO_2 equivalent units, perhaps another 50–70 CO_{2e} making the number something closer to 500 ppm.

3. Intergovernmental Panel on Climate Change, *Climate Change 2022: Impacts, Adaptation and Vulnerability* (February 2022); and Mark Lynas, *Our Final Warning* (London: 4th Estate, 2020); How much additional warming in the “pipeline” is still to be determined. See James Hansen et.al., “Global warming in the pipeline,” unpublished manuscript under review. December 13, 2022.

4. Timothy Lenton et al., “Climate Tipping Points—Too Risky to Bet Against,” *Nature* 27 (November 2019).

5. H. Damon Matthews and Seth Wynes, “Current Global Efforts Are Insufficient to Limit Warming to 1.5°C,” *Science* 376, no. 6600 (2022): 1404–1409; Luke Kemp et al., “Climate Endgame: Exploring Catastrophic Climate Change Scenarios,” *Proceedings of the National Academy of Sciences* 119, no. 34 (2022); David Armstrong et al., “Exceeding 1.5°C Global Warming Could Trigger Multiple Climate Tipping Points,” *Science* 9 (September 2022).

6. Naomi Klein, *This Changes Everything* (New York: Simon & Schuster, 2014).

7. “But if this is the last step,” Yale economist William Nordhaus said, “then we are in for a fiery future,” quoted in Coral Davenport and Lisa Freeman, “Five Decades in the Making,” *New York Times*, August 8, 2022; also Jeffrey Sachs, https://www.commondreams.org/views/2022/08/20/four-reasons-democrats-inflation-reduction-act-doesnt-live-hype.

8. Mark Z. Jacobson, “Low-Cost Solutions to Global Warming, Air Pollution, and Energy Insecurity,” *Energy and Environmental Science* (Royal Society of Chemistry) (June 2022); Amory Lovins, “Interview,” in *Energy Intelligence*, May 27, 2022.

9. But, as Alex Connon reports, "Wall Street's financing of coal, oil, and gas was higher in 2021 than it was in 2016 and US banks provided $64 bn in financing to corporations most rapidly expanding coal, oil, and gas," *Manchester Guardian,* July 28, 2022.

10. Delay means, among other things, that the annual percentage change necessary to stay below 1.5°C rises each year. It is now estimated to be over 8 percent/year and rising. See Stan Cox.

11. Gus Speth, *They Knew* (Cambridge: MIT Press, 2021), 5.

12. Paul Hawken et al. catalogued and analyzed the various alternatives to our present course in *Drawdown: The Most Comprehensive Plan Ever Proposed to Reverse Global Warming* (New York: Penguin, 2017).

13. James Hansen, Makiko Sato, and Reto Ruedy, "June 2022 Temperature Update & the Bigger Picture," email, July 29, 2022; David Archer, in *The Long Thaw* (Princeton, NJ: Princeton University Press, 2009) writes that a quarter of CO_2 released now will "still be affecting the climate one thousand years from now." In an email on December 14, 2021, he estimates the time required to stabilize the climate at "hundreds of millennia." As for mechanical ways to remove CO_2, the challenge is daunting: it would itself have to be carbon neutral, affordable, deployed at a scale to remove ~nine billion tons/year, and at a pace fast enough to prevent the worst outcomes. And what do we do with the carbon captured each year, estimated to be equivalent in volume to that presently transported in oil/gas pipelines?

14. Wendell Berry, "Solving for Pattern," in Berry, *The Gift of Good Land* (San Francisco: North Point Press, 1981), 134–145.

15. John Dryzek, Richard Norgaard, David Schlosberg, "Climate Change and Society," in *The Oxford Handbook of Climate Change and Society*, ed. Dryzek et al. (New York: Oxford University Press, 2011), 3–17.

16. V-Dem Institute, *Autocratization Changing Nature?* (V-Dem Institute: University of Gothenburg, 2022); Max Fisher, "How Democracy Is Under Threat across the Globe," August 19, 2022, https://www.nytimes.com/2022/08/19/world/democracy-threat.html?smid=url-share.

17. Jason Brennan and Hélène Landemore, *Debating Democracy* (New York: Oxford University Press, 2022), 180–210.

18. See Richard Rorty's concise defense of democracy in more straitened times in Richard Rorty, "Rethinking Democracy (1996)," in Rorty, *What Can We Hope For?* (Princeton, NJ: Princeton University Press, 2022), 65–70.

19. Amitav Ghosh, *The Great Derangement* (Chicago: University of Chicago Press, 2016), 160.

20. Walter Prescott Webb, *The Great Frontier* (Austin: University of Texas Press, 1964 [1951]); but the practice of democracy is old and long preceded the Western version of democracy, see David Stasavage, *The Decline and Rise of Democracy* (Princeton, NJ: Princeton University Press, 2020).

21. Michael Klare, *The Race for What's Left* (New York: Metropolitan Books, 2012).

22. https://www.reuters.com/investigates/special-report/assets/usa-courts-secrecy-lobbyist/powell-memo.pdf; Gary Gerstle, *The Rise and Fall of the Neoliberal Order* (New York: Oxford University Press, 2022), 108–140; Robert Kuttner, "Free Markets, Besieged Citizens," *New York Review of Books*, July 21, 2022, pp. 12–14; Naomi Oreskes and Erik M. Conway, *The Big Myth: How American Business Taught Us to Loath Government and Love the Free Market* (New York: Bloomsbury, 2023).

23. Enshrined in the 2010 Supreme Court decision, Citizens United v. Federal Election Commission.

24. "It's worse than you think," is likely true because the commitment to democracy is not equally shared. Dana Milbank, *The Destructionists* (New York: Doubleday, 2022); Nicole Hemmer, *Partisans* (New York: Basic Books, 2022).

25. Katherine Stewart, *The Power Worshippers* (New York: Bloomsbury, 2019); and Stewart, "Christian Nationalists Are Excited about What Comes Next," *New York Times*, July 5, 2022. "It's purpose," she writes, "is to hollow out democracy until nothing is left but a thin cover for rule by a supposedly right-thinking elite, bubble-wrapped in sanctimony and insulated from any real democratic check on its power." Amen.

26. Rogers et al., "Sixfold Increase in Historical Northern Hemisphere Concurrent Large Heatwaves Driven by Warming and Changing Atmospheric Circulations," *Journal of Climate* 35, no. 3 (2022): 1063–1078; and on decreased civility, see Annika Stechemesser, Anders Levermann, and Leonie Wenz, "Temperature Impacts on Hate Speech Online," *The Lancet* 5, no. 9 (2022): E714–E725.

27. My use of the word "healing" democracy owes a great deal to Parker Palmer's *Healing the Heart of Democracy* (San Francisco: Jossey-Bass, 2011).

28. Denise Fairchild and Al Weinrub, eds., *Energy Democracy* (Washington, DC: Island Press, 2017).

29. W. E. B. Du Bois, *Dark Water: Voices from within the Veil* (New York: Dover, [1920] 1999), 84.

30. David W. Orr, "The Political Economy of Design in a Hotter Time," in *Routledge Handbook of Sustainable Design*, ed. Rachel Berth Egenhoefer (London: Routledge, 2018), 3–9.

31. John Wesley Powell, *Report on the Lands and Arid Region of the United States* (Boston: Harvard Commons Press, 1983 [1879]); Herman E. Daly and John B. Cobb, *For the Common Good* (Boston: Beacon Press, 1994); Herman E. Daly and Joshua Farley, *Ecological Economics* (Washington, DC: Island Press, 2011); Thomas Homer-Dixon, *The Ingenuity Gap* (New York: Knopf, 2000).

32. "This expansion cannot go on forever and ultimately we must face some vexatious issues of social justice." Reinhold Niebuhr, *The Irony of American History* (New York: Scribner's, 1952), 29.

33. Vaclav Havel, *Living in Truth* (London: Faber and Faber, 1989); also Tony Judt, *Ill Fares the Land* (New York: Penguin Press, 2010), 1–9.

34. Robert Sapolsky, *Behave: The Biology of Humans at Our Best and Worst* (New York: Penguin, 2017); Christopher Lasch, *The Culture of Narcissism* (New York: Norton, 1979); M. Scott Peck, *People of the Lie* (New York: Simon and Schuster, 1983), particularly pp. 212–269.

35. On the role of capitalism, I agree with Nancy Fraser that climate change is not an anomaly, but rather the logical working out of the rules of capitalism as presently conceived. See Nancy Fraser, *Cannibal Capitalism* (London: Verso, 2022), 85.

36. James Gustave Speth, "Preface," in *The New Systems Reader*, ed. James Gustave Speth and Kathleen Courrier (New York: Routledge, 2021), xx–xxvi; Donella Meadows, *Thinking in Systems* (White River Junction, VT: Chelsea Green, 2008).

37. Klein, *This Changes Everything*, 22.

THE CHALLENGE

1. Spencer R. Weart, "How Could Climate Change," in *The Discovery of Global Warming* (Cambridge, MA: Harvard University Press, 2008), chapter 1.

2. R. Revelle and H. E. Seuss, "Carbon Dioxide Exchange between Atmosphere and Ocean and the Question of an Increase in Atmospheric CO_2 during the Past Decades," *Tellus* 9, no. 1 (1957): 18–27.

3. United Nations Framework Convention on Climate Change, unfccc.int, 1992, https://unfccc.int/resource/docs/convkp/conveng.pdf.

4. United Nations Framework Convention on Climate Change, The Paris Agreement, unfccc.int, https://unfccc.int/sites/default/files/resource/parisagreement_publication.pdf.

5. IPCC, "2021: Summary for Policymakers," in *Climate Change 2021: The Physical Science Basis*, Contribution of Working Group I to the Sixth Assessment Report of the Intergovernmental Panel on Climate Change, ed. V. Masson-Delmotte et al. (Cambridge, UK: Cambridge University Press), 3–32, https://doi.org/10.1017/9781009157896.001, figure SPM.8, p. 22; https://www.ipcc.ch/report/ar6/wg1/downloads/report/IPCC_AR6_WGI_SPM.pdf.

6. IPCC, "2021: Summary for Policymakers," in *Climate Change 2021*, 3–32, https://doi.org/10.1017/9781009157896.001, p. 11; https://www.ipcc.ch/report/ar6/wg1/downloads/report/IPCC_AR6_WGI_SPM.pdf.

7. IPCC, "2021: Summary for Policymakers," in *Climate Change 2021*, 3–32, https://doi.org/10.1017/9781009157896.001, Figure SPM.6, p. 18; https://www.ipcc.ch/report/ar6/wg1/downloads/report/IPCC_AR6_WGI_SPM.pdf.

8. "2021 Western North America Heat Wave," Wikipedia, edited September 9, 2022, https://en.wikipedia.org/wiki/2021_Western_North_America_heat_wave.

9. M. Oppenheimer et al., "Sea Level Rise and Implications for Low-Lying Islands, Coasts and Communities," in *IPCC Special Report on the Ocean and Cryosphere in a Changing Climate,* ed. H.-O. Pörtner et al., fig. 4.12, 359, ipcc.ch, 2019, https://www.ipcc.ch/site/assets/uploads/sites/3/2019/11/08_SROCC_Ch04_FINAL.pdf.

10. IPCC, "2021: Summary for Policymakers," in *Climate Change 2021*, 3–32, https://doi.org/10.1017/9781009157896.001, Figure SPM.6, p. 18; https://www.ipcc.ch/report/ar6/wg1/downloads/report/IPCC_AR6_WGI_SPM.pdf.

11. R. McLeman et al., "Conceptual Framing to Link Climate Risk Assessments and Climate-Migration Scholarship," *Climatic Change* 165, no. 24 (March 2021), https://doi.org/10.1007/s10584-021-03056-6.

12. H. Benveniste, M. Oppenheimer, and M. Fleurbaey, "Effect of Border Policy on Exposure and Vulnerability to Climate Change," *Proceedings of the National Academy of Sciences* 117 (October 2020): 26692–26702, https://www.pnas.org/cgi/doi/10.1073/pnas.2007597117.

CHAPTER 1

1. K. C. Krishna Bahadur et al., "When Too Much Isn't Enough: Does Current Food Production Meet Global Nutritional Needs?," *PLOS One* 13, no. 10 (October 2018): e0205683, https://doi.org/10.1371/journal.pone.0205683.

2. "Three Billion People Cannot Afford Healthy Diets: What Does This Mean for the Next Green Revolution?," Center for Strategic and International Studies, September 23, 2020, https://www.csis.org/analysis/three-billion-people-cannot-afford-healthy-diets-what-does-mean-next-green-revolution.

3. Hannah Ritchie, "How Much of the World's Land Would We Need in Order to Feed the Global Population with the Average Diet of a Given Country?," Our World in Data, October 3, 2017, https://ourworldindata.org/agricultural-land-by-global-diets.

4. Krista Charles, "Food Production Emissions Make Up More Than a Third of Global Total," *New Scientist*, September 13, 2021, https://www.newscientist.com/article/2290068-food-production-emissions-make-up-more-than-a-third-of-global-total/; Tim G. Benton et al., "Food System Impacts on Biodiversity Loss," Chatham House, February 2021, https://www.chathamhouse.org/sites/default/files/2021-02/2021-02-03-food-system-biodiversity-loss-benton-et-al_0.pdf; Jenny Howard, "Dead Zones, Facts and Information," *National Geographic*, July 31, 2019, https://www.nationalgeographic.com/environment/article/dead-zones; Robert W. Howarth et al., "Nitrogen Use in the United States from 1961–2000 and Potential Future Trends," *Ambio* 31, no. 2 (2002): 88–96, https://www.jstor.org/stable/4315220.

5. Howard, "Dead Zones," https://www.nationalgeographic.com/environment/article/dead-zones; Jonathan Watts, "1% of Farms Operate 70% of World's Farmland," *Guardian*, November 24, 2020, https://www.theguardian.com/environment/2020/nov/24/farmland-inequality-is-rising-around-the-world-finds-report.

6. Sarah Repucci and Amy Slipowitz, "The Global Expansion of Authoritarian Rule," Freedom House, February 2022, https://freedomhouse.org/report/freedom-world/2022/global-expansion-authoritarian-rule.

7. Albert Einstein quote, "It is the theory that decides what can be observed," AZ Quotes, accessed September 7, 2022, https://www.azquotes.com/quote/531990.

8. Plato quote, "Those who tell the stories rule society," Quote Fancy, accessed July 6, 2022, https://quotefancy.com/quote/13855/Plato-Those-who-tell-the-stories-rule-society; Glenn Leibowitz, "Those Who Tell the Stories Rule the World," *Write with Impact* (blog), August 16,

2015, https://medium.com/write-with-impact/those-who-tell-the-stories-rule-the-world-8129bd07090c.

9. "About," Inequality.org, accessed June 24, 2022, https://inequality.org/about/.

10. Juliana Menasce Horowitz, Ruth Igielnik, and Rakesh Kochhar, "1. Trends in Income and Wealth Inequality," Pew Research Center, January 9, 2020, https://www.pewresearch.org/social-trends/2020/01/09/trends-in-income-and-wealth-inequality/.

11. "Gini Index | Data," World Bank, accessed September 7, 2022, https://data.worldbank.org/indicator/SI.POV.GINI?most_recent_value_desc=false.

12. "The Economic Cost of Food Monopolies: The Grocery Cartels," Food and Water Watch, November 2021, https://www.foodandwaterwatch.org/wp-content/uploads/2021/11/IB_2111_FoodMonoSeries1-SUPERMARKETS.pdf; Reuters, "Explainer: How Four Big Companies Control the U.S. Beef Industry," *Reuters*, June 17, 2021, sec. Business, https://www.reuters.com/business/how-four-big-companies-control-us-beef-industry-2021-06-17/.

13. USDA ERS, "Farmland Ownership and Tenure," United States Department of Agriculture, May 16, 2022, https://www.ers.usda.gov/topics/farm-economy/land-use-land-value-tenure/farmland-ownership-and-tenure/.

14. USDA NASS, "Farms and Farmland," United States Department of Agriculture, August 2019, https://www.nass.usda.gov/Publications/Highlights/2019/2017Census_Farms_Farmland.pdf.

15. Nick Estes, "Bill Gates Is the Biggest Private Owner of Farmland in the United States. Why?" *Guardian*, April 5, 2021, sec. Opinion, https://www.theguardian.com/commentisfree/2021/apr/05/bill-gates-climate-crisis-farmland; USDA ERS, "Farming and Farm Income," United States Department of Agriculture, September 1, 2022, https://www.ers.usda.gov/data-products/ag-and-food-statistics-charting-the-essentials/farming-and-farm-income/.

16. Farm Aid, "Corporate Control of Agriculture," accessed July 13, 2022, https://www.farmaid.org/issues/corporate-power/corporate-power-in-ag/.

17. Nancy Fink Huehnergarth, "Nearly 60% of Calories, 90% of Added Sugars in U.S. Diet Are from 'Ultra-Processed' Foods," *Forbes*, March 14, 2018, https://www.forbes.com/sites/nancyhuehnergarth/2016/03/14/study-finds-nearly-60-of-calories-in-the-american-diet-are-from-ultra-processed-foods/.

18. "Cancer: Carcinogenicity of the Consumption of Red Meat and Processed Meat," World Health Organization, October 26, 2015, https://www.who.int/news-room/questions-and-answers/item/cancer-carcinogenicity-of-the-consumption-of-red-meat-and-processed-meat.

19. Cheikh Mbow and Cynthia Rosenzweig, "Chapter 5: Food Security—Special Report on Climate Change and Land," in *Climate Change and Land: An IPCC Special Report* (Geneva: IPCC, 2019), https://www.ipcc.ch/srccl/chapter/chapter-5/.

20. "What Are the Biggest Drivers of Tropical Deforestation?," *World Wildlife Magazine* (Summer 2018), https://www.worldwildlife.org/magazine/issues/summer-2018/articles/what-are-the-biggest-drivers-of-tropical-deforestation.

21. "Influence of Big Money," Brennan Center for Justice, accessed July 6, 2022, https://www.brennancenter.org/issues/reform-money-politics/influence-big-money.

22. Robert D. McFadden, "David Koch, Billionaire Who Fueled Right-Wing Movement, Dies at 79," *New York Times*, August 23, 2019, sec. US, https://www.nytimes.com/2019/08/23/us/david-koch-dead.html.

23. Joseph Zeballos-Roig, "How the Koch Brothers Used Their Massive Fortune to Power a Conservative Crusade That Reshaped American Politics," *Business Insider*, November 13, 2020, https://www.businessinsider.com/koch-brothers-fortune-power-conservative-crusade-american-politics-2019-8.

24. David M. Drucker, "Koch Network Group Plans Big-Time, Pro-GOP 2022 Candidate Investments," *Washington Examiner*, April 19, 2022, https://www.washingtonexaminer.com/news/campaigns/koch-network-group-plans-big-time-pro-gop-2022-candidate-campaign-investment.

25. "Politicians & Elections," overview," OpenSecrets, accessed July 6, 2022, https://www.opensecrets.org/elections/.

26. Dick Dahl, "Fear and Loathing," *Harvard Law Today*, December 6, 2012, https://today.law.harvard.edu/book-review/in-new-book-lessig-diagnoses-a-cancer-that-has-attacked-our-political-system/.

27. "Lobbying Data Summary," OpenSecrets, April 22, 2022, https://www.opensecrets.org/federal-lobbying.

28. "Countries and Territories," Freedom House, accessed June 24, 2022, https://freedomhouse.org/countries/freedom-world/scores; "Our History," Freedom House, accessed June 24, 2022, https://freedomhouse.org/about-us/our-history.

29. Sarah F. Brosnan and Frans B. M. de Waal, "Monkeys Reject Unequal Pay," *Nature*, September 18, 2003, https://www.emory.edu/LIVING_LINKS/publications/articles/Brosnan_deWaal_2003.pdf.

30. Adam Smith, *The Theory of Moral Sentiments* (Amsterdam: MetaLibri, 1970), https://www.adamsmith.org/the-theory-of-moral-sentiments.

31. Ágnes Melinda Kovács, Ernő Téglás, and Ansgar Denis Endress, "The Social Sense: Susceptibility to Others' Beliefs in Human Infants and Adults," *Science* 330, no. 6012 (2010): 1830–1834, https://doi.org/10.1126/science.1190792.

32. M. Dondi, F. Simion, and G. Caltran, "Can Newborns Discriminate between Their Own Cry and the Cry of Another Newborn Infant?" *Developmental Psychology* 35, no. 2 (1999): 418–426, https://doi.org/10.1037//0012-1649.35.2.418.

33. Jules H. Masserman, Stanley Wechkin, and Terris William, "'Altruistic' Behavior in Rhesus Monkeys," *American Journal of Psychiatry* 121 (December 1964): 584–585, http://www.primatefreedom.com/masserman.pdf.

34. John Drury and Stephen D. Reicher, "Crowd Control," *Scientific American Mind* (November 2010).

35. "The Secret to Happiness? Giving," March 20, 2008, *ScienceNews*, https://www.science.org/content/article/secret-happiness-giving.

36. Democracy Movement (www.democracymovement.us) is cosponsored by the Small Planet Institute and Democracy Initiative, a national coalition of approximately forty-five social-benefit organizations across many issues that also engage in advancing democracy reforms. The Democracy Movement website enables citizens in every state to find nearby democracy initiatives with which to engage.

37. "What Is Meaning of Eleanor Roosevelt's Quote 'Do One Thing Every Day That Scares You'?" Quora, accessed September 7, 2022, https://www.quora.com/What-is-meaning-of-Eleanor-Roosevelts-quote-Do-one-thing-every-day-that-scares-you.

38. "What Is Democracy?," Facing History and Ourselves, November 15, 2017, https://www.facinghistory.org/resource-library/what-democracy.

CHAPTER 2

1. Michael Mendez, *Climate Change from the Streets* (New Haven: Yale university Press, 2020), 30.

2. These additions in part reflected the presence and influence of representatives from the French ultramarine territories, which represent 70 percent of French biodiversity and whose population has been poisoned by a carcinogenic pesticide known as chlordecone.

3. Sciences-po, poll Baromètre de la confiance politique, p. 132, https://www.sciencespo.fr/cevipof/sites/sciencespo.fr.cevipof/files/OpinionWay%20pour%20le%20CEVIPOF-Baromètre%20de%20la%20confiance%20en%20politique%20-%20vague12%20-%20Rapport%20international%20(1).pdf.

4. See English translation of the 2011 constitutional proposal, http://stjornlagarad.is/other_files/stjornlagarad/Frumvarp-enska.pdf.

5. https://www.as-coa.org/articles/look-what-and-isnt-chiles-constitutional-draft.

6. All results are available here: http://www.odoxa.fr/sondage/mesures-de-convention-citoyenne-seduisent-francais-a-lexception-notable-110-km-h/].

7. Elabe survey for the Climate Action Network on June 23 and 24, 2020, Convention citoyenne pour le climat, qu'en pensent les Français?, https://reseauactionclimat.org/sondage-des-gaulois-pas-si-refractaires-a-laction-climatique/. An Odoxa survey for Le Figaro and France Info on June 24 and 25, 2020, La Convention citoyenne pour le climat, also indicated that four out of five French people were in favor of putting the main measures to a referendum, http://www.odoxa.fr/sondage/mesures-de-convention-citoyenne-seduisent-francais-a-lexception-notable-110-km-h.

8. Elizabeth Anderson, *Private Government: How Employers Rule Our Lives (and Why We Don't Talk about It)* (Princeton, NJ: Princeton University Press, 2017).

9. Such as that proposed by Pavlivna Tcherneva, *The Case for a Job Guarantee* (Cambridge, UK: Polity, 2020).

10. See Isabelle Ferreras, *Firms as Political Entities: Saving Democracy through Economic Bicameralism* (Cambridge, UK: Cambridge University Press, 2017).

11. Isabelle Ferreras, Julie Battilana, and Dominique Méda, *Democratize Work* (Chicago: Chicago University Press, 2022). Full disclosure: I'm a contributor to that book and larger movement.

CHAPTER 3

1. Alexis de Tocqueville, *Democracy in America* (Chicago: University of Chicago Press, [1835] 2000).

2. Francis Fukuyama, *The End of History and the Last Man* (New York: Free Press, [1992] 2006).

3. Al Gore, *Earth in the Balance: Ecology and the Human Spirit* (Boston: Houghton Mifflin, 1992).

4. Marina Povitkina and Sverker C. Jagers, "Environmental Commitments in Different Types of Democracies: The Role of Liberal, Social-Liberal, and Deliberative Politics," V-Dem Working Paper 117, March 1, 2021, https://papers.ssrn.com/sol3/papers.cfm?abstract_id=3810624.

5. Torbjørn Selseng, Kristin Linnerud, and Erling Holden, "Unpacking Democracy: The Effects of Different Democratic Qualities on Climate Change Performance Over Time," *Environmental Science and Policy* 128 (2022).

6. O. M. Lægreid and M. Povitkina, "Do Political Institutions Moderate the GDP-CO2 Relationship?" *Ecological Economics* 145 (2018): 441–450.

7. United Nations Environment Programme, "Emissions Gap Report 2021: The Heat Is On—A World of Climate Promises Not Yet Delivered," https://wedocs.unep.org/20.500.11822/36990.

8. D. Gilbert, *Stumbling on Happiness* (New York: Knopf Doubleday, 2006).

9. I. Lorenzoni, S. Day, and L. Whitmarsh, "Barriers Perceived to Engaging with Climate Change among the UK Public and Their Policy Implications," *Global Environmental Change* 17 (2007): 445–459.

10. K. Schaeffer, "State of the Union 2022: How Americans View Major National Issues," Pew Research Center, 2022, https://www.pewresearch.org/fact-tank/2022/02/25/state-of-the-union-2022-how-americans-view-major-national-issues/.

11. R. Hoffmann, R. Muttarak, J. Peisker, and P. Stanig, "Climate Change Experiences Raise Environmental Concerns and Promote Green Voting," *Nature Climate Change* (2022).

12. S. Carattini et al., "Green Taxes in a Post-Paris World: Are Millions of Nays Inevitable?" *Environmental and Resource Economics* 68 (2017): 97–128.

13. Statistica, "Voter Turnout Rates among Selected Age Groups in U.S. Presidential Elections from 1964–2020," 2022, https://www.statista.com/statistics/1096299/voter-turnout-presidential-elections-by-age-historical/.

14. M. Karlsson, E. Alfredsson, and N. Westling, "Climate Policy Co-benefits: A Review," *Climate Policy* 20, no. 3 (2020): 292–316.

15. S. Carattini, S. Kallbekken, and A. Orlov, "How to Win Public Support for a Global Carbon Tax," *Nature* 565 (2019): 289–291.

16. L. C. Hamilton, J. Hartter, and E. Bell, "Generation Gaps in US Public Opinion on Renewable Energy and Climate Change," *PLOS One* 14, no. 7 (2019): e0217608, https://doi.org/10.1371/journal.pone.0217608.

17. R. Brulle, "The Climate Lobby: A Sectoral Analysis of Lobbying Spending on Climate Change in the USA, 2000–2016," *Climatic Change* 149, no. 3/4 (2018): 289–303.

18. M. Povitkina, "The Limits of Democracy in Tackling Climate Change," *Environmental Politics* 27, no. 3 (2018): 411–432.

19. "Grantham Research Institute on Climate Change and the Environment and Sabin Center for Climate Change Law," Climate Change Laws of the World database, 2022, https://climate-laws.org/.

20. S. Eskander, and S. Fankhauser, "Reduction in Greenhouse Gas Emissions from National Climate Legislation," *Nature Climate Change* 10 (2020): 750–756.

21. World Bank, "Carbon Pricing Dashboard," 2022, https://carbonpricingdashboard.worldbank.org/map_data.

22. R. Best, P. J. Burke, and F. Jotzo, "Carbon Pricing Efficacy: Cross-Country Evidence," *Environmental and Resource Economics* 77 (2020): 69–94.

23. R. O'Gorman, "Environmental Constitutionalism: A Comparative Study," *Transnational Environmental Law* 6, no. 3 (2017): 435–462.

24. K. Toral et al., "The 11 Nations Heralding a New Dawn of Climate Constitutionalism," 2021, https://www.lse.ac.uk/granthaminstitute/news/the-11-nations-heralding-a-new-dawn-of-climate-constitutionalism/.

25. L. Beckman, "Democracy and Future Generations: Should the Unborn Have a Voice?" In *Spheres of Global Justice: Vol. 2: Fair Distribution-Global Economic, Social and Intergenerational Justice*, ed. J. C. Merle (Dordrecht: Springer, 2013), 775–788.

26. L. Bronner and D. Ifkovits, "Voting at 16: Intended and Unintended Consequences of Austria's Electoral Reform," *Electoral Studies* 61 (October 2019).

27. M. Miljand and K. Bäckstrand, "Climate Policy Councils: Success Factors and Lessons Learned," *A GCF-CEEW Report* (Stockholm: Global Challenges Foundation, 2021).

28. J. Setzer and C. Higham, "Global Trends in Climate Change Litigation: 2021 Snapshot," Grantham Research Institute on Climate Change and the Environment and Centre for Climate Change Economics and Policy (London: London School of Economics and Political Science, 2021).

29. J. Peel and H. M. Osofsky, "Climate Change Litigation (October 2020)," *Annual Review of Law and Social Science* 16 (2020): 21–38.

30. J. Krommendijk, "Beyond Urgenda: The Role of the ECHR and Judgments of the ECtHR in Dutch Environmental and Climate Litigation," *Review of European, Comparative and International Environmental Law*, June 30, 2021, https://doi.org/10.1111/reel.12405.

31. Intergovernmental Panel on Climate Change, "The Sixth Assessment Report of the IPCC on Mitigation of Climate Change," Working Group III Contribution to the Sixth Assessment Report of the Intergovernmental Panel on Climate Change (2022).

32. K. Pouikli, "Editorial: A Short History of the Climate Change Litigation Boom across Europe," *ERA Forum* 22 (2022).

33. L. Burgers, "Should Judges Make Climate Change Law?" *Transnational Environmental Law* 9, no. 1 (2020): 55–75.

CHAPTER 4

1. Langston Hughes, "Let America be America Again," 1935.

2. Rev. Martin Luther King, "Beyond Vietnam: A Time to Break Silence," 4 April 1967, https://inside.sfuhs.org/dept/history/US_History_reader/Chapter14/MLKriverside.htm.

3. Sarah Anderson, Marc Bayard, and Phyllis Bennis, "The Souls of Poor Folk: Auditing America 50 Years after the Poor People's Campaign Challenged Racism, Poverty, the War Economy/Militarism, and Our National Morality," Institute for Policy Studies, April 2018, https://www.poorpeoplescampaign.org/wp-content/uploads/2019/12/PPC-Audit-Full-410835a.pdf.

4. Repairers of the Breach, Kairos Center, and Poor Peoples Campaign, *A Moral Policy Agenda to Heal and Transform America*, July 2020, https://www.poorpeoplescampaign.org/wp-content/uploads/2020/08/PPC-Policy-Platform_8-28.pdf.

5. Dr. Martin Luther King, "Where Do We Go from Here," August 16, 1967, http://www.beacon.org/The-Radical-King-P1166.aspx.

6. Bill D. Moyers, "What a Real President Was Like," *Washington Post*, November 13, 1988, https://www.washingtonpost.com/archive/opinions/1988/11/13/what-a-real-president-was-like/d483c1be-d0da-43b7-bde6-04e10106ff6c/.

7. Du Bois, W. E. B., *Black Reconstruction in America: Toward a History of the Part Which Black Folk Played in the Attempt to Reconstruct Democracy in America*, 1860–1880 (New York: Routledge, 2012).

8. Ibram Kendi, "What to an American Is the Fourth of July?" *The Atlantic*, July 4, 2019.

9. Philip Alston, "Report of the Special Rapporteur on Extreme Poverty and Human Rights," June 25, 2019.

10. Ovais Sarmad, "2020 a 'Critical Year for Addressing Climate Change'," January 23, 2020, https://unfccc.int/news/2020-a-critical-year-for-addressing-climate-change-ovais-sarmad

11. D. J. X. Gonzalez et al., "Historic Redlining and the Siting of Oil and Gas Wells in the United States, *Journal of Exposure Science and Environmental Epidemiology* (2022), https://doi.org/10.1038/s41370-022-00434-9.

12. Federal Housing Administration, Underwriting Manual: Underwriting and Valuation Procedure Under Title II of the National Housing Act With Revisions to April 1, 1936 (Washington, D.C.), Part II, Section 2.

13. Richard Rothstein, *The Color of Law* (New York: Liveright, 2018).

14. Van Jones, *The Green-Collar Economy: How One Solution Can Fix Our Two Biggest Problems* (New York: HarperOne, 2008)

15. Dolf Gielen, "IRENA World Energy Transitions Outlook 1.5°C Pathway," International Renewable Energy Agency, July 8, 2021, Conference: SAIFAC Climate Change Leadership Seminar Series, https://www.researchgate.net/publication/353325586_World_Energy_Transitions_Outlook_15_C_Pathway.

CHAPTER 5

1. Available at www.genspace.org/mission.

2. Detlev Bronk, "The National Science Foundation: Origins, Hopes, and Aspirations," *Science* 188, 4187 (1975): 409–414.

3. Robert Dahl, *Controlling Nuclear Weapons: Democracy versus Guardianship* (Syracuse: Syracuse University Press, 1985), 6.

CHAPTER 6

1. "Human Rights Foundation," accessed June 26, 2022, https://hrf.org/about/mission/.

2. David T. Courtwright, *The Age of Addiction: How Bad Habits Became Big Business* (Cambridge, MA: Belknap Press/Harvard University Press, 2019).

3. Tressie McMillan Cottom, "Where Platform Capitalism and Racial Capitalism Meet: The Sociology of Race and Racism in the Digital Society," *Sociology of Race and Ethnicity* 6, no. 4 (October 2020): 441–449, https://doi.org/10.1177/2332649220949473.

4. A. Marwick, R. Kuo, S. J. Cameron, and M. Weigel, "Critical Disinformation Studies: A Syllabus," Center for Information, Technology, and Public Life (CITAP), University of North Carolina at Chapel Hill, nd, http://citap.unc.edu/critical-disinfo.

5. Nathaniel Persily and Joshua Tucker, eds., *Social Media and Democracy: The State of the Field, Prospects for Reform*, SSRC Anxieties of Democracy (Cambridge, UK: Cambridge University Press, 2020).

6. Geert Lovink, "This Level of Metaphysics Ain't No Shit: Geert Lovink Interviewed by Bram Ieven," *3:AM Magazine*, 2020, https://www.3ammagazine.com/3am/3am-in-lockdown-26-geert-lovink/.

7. Imre Szeman and Caleb Wellum, "Carbon Democracy at Ten: An Interview with Timothy Mitchell," *Cultural Studies*, March 30, 2022, pp. 1–19, https://doi.org/10.1080/09502386.2022.2056221.

8. Bregie van Veelen and Dan van der Horst, "What Is Energy Democracy? Connecting Social Science Energy Research and Political Theory," *Energy Research and Social Science* 46 (December 2018): 19–28, https://doi.org/10.1016/j.erss.2018.06.010.

9. Hillary Aidun et al., "Opposition to Renewable Energy Facilities in the United States: March 2022 Edition," nd, 96.

10. "The Role of Critical Minerals in Clean Energy Transitions," nd, 287.

11. FAO, *The State of Food Security and Nutrition in the World 2022*, 2022, https://doi.org/10.4060/cc0639en.

12. Jasmien De Winne and Gert Peersman, "The Impact of Food Prices on Conflict Revisited," *Journal of Business and Economic Statistics* 39, no. 2 (April 2021): 547–560, https://doi.org/10.1080/07350015.2019.1684301.

13. Katherine J. Mach et al., "Climate as a Risk Factor for Armed Conflict," *Nature* 571, no. 7764 (July 2019): 193–197, https://doi.org/10.1038/s41586-019-1300-6.

14. IPCC, "Climate Change 2022: Impacts, Adaptation and Vulnerability, Summary for Policymakers," nd.

15. Bronislaw Szerszynski et al., "Why Solar Radiation Management Geoengineering and Democracy Won't Mix," *Environment and Planning A: Economy and Space* 45, no. 12 (December 2013): 2809–2816, https://doi.org/10.1068/a45649.

16. Joshua B. Horton et al., "Solar Geoengineering and Democracy," *Global Environmental Politics* 18, no. 3 (August 2018): 5–24, https://doi.org/10.1162/glep_a_00466.

17. Geoff Mann and Joel Wainwright, *Climate Leviathan: A Political Theory of Our Planetary Future* (London: Verso, 2018).

18. Ben Tarnoff, *Internet for the People: The Fight for Our Digital Future* (London: Verso, 2022).

CHAPTER 7

1. Frederick Mayer and Gary Gereffi, "Regulation and Economic Globalization: Prospects and Limits of Private Governance," *Business and Politics* 12, no. 3 (2010): 1–25.

2. United Nations Environment Programme, "Adaptation Gap Report 2021: The Gathering Storm—Adapting to Climate Change in a Post Covid World," 2021, https://wedocs.unep.org/xmlui/bitstream/handle/20.500.11822/37312/AGR21_ESEN.pdf.

3. There is a considerable literature on multilevel negotiations. The earliest works are Frederick Mayer, "Bargains within Bargains: Domestic Politics and International Relations," PhD diss., Harvard University (1988); and Robert D. Putnam, "Diplomacy and Domestic Politics: The Logic of Two-Level Games," *International Organization* 42, no. 3 (1988): 427–460.

4. Institute for Democracy and Electoral Assistance, "The Global State of Democracy Report 2021: Building Resilience in a Pandemic Era" (Strömsborg, Sweden: International IDEA), https://www.idea.int/gsod/.

5. Daniel Lindvall, "Democracy and the Challenge of Climate Change," International IDEA Discussion Paper 3/2021, Institute for Democracy and Electoral Assistance, 2021, https://www.idea.int/sites/default/files/publications/democracy-and-the-challenge-of-the-climate-change.pdf.

6. Pew Research Center, "Global Public Opinion in an Era of Democratic Anxiety," 2021, https://www.pewresearch.org/global/2021/12/07/global-public-opinion-in-an-era-of-democratic-anxiety/

7. Robert O. Keohane and David G. Victor, "The Regime Complex for Climate Change," *Perspectives on Politics* 9, no. 1 (2011): 7–23.

8. Jennifer Marlon et al., *Yale Climate Opinion Maps 2021*, Yale Program on Climate Change Communication, https://climatecommunication.yale.edu/visualizations-data/ycom-us/.

9. A. Leiserowitz et al., *International Public Opinion on Climate Change, 2022* (New Haven, CT: Yale Program on Climate Change Communication and Data for Good at Meta, 2022), https://climatecommunication.yale.edu/wp-content/uploads/2022/06/international-public-opinion-on-climate-change-2022a.pdf.

10. United Nations Development Programme, *People's Climate Vote*, 2021, https://www.undp.org/publications/peoples-climate-vote.

11. Pew Research Center, "Global Public Opinion in an Era of Democratic Anxiety," 2021, https://www.pewresearch.org/global/2021/12/07/global-public-opinion-in-an-era-of-democratic-anxiety/.

12. Frederick W. Mayer, *Narrative Politics: Stories and Collective Action* (New York: Oxford University Press, 2014).

CHAPTER 8

1. Luke Kemp et al., "Climate Endgame: Exploring Catastrophic Climate Change Scenarios," *Proceedings of the National Academy of Sciences*, August 2022.

2. Ryan Heath, "The World Is on Fire, and Our Leaders Are Failing," *Politico*, August 2022.

3. International Energy Agency, "Net Zero by 2050: A Roadmap for the Global Energy Sector," May 2021. https://www.iea.org/reports/net-zero- by-2050.

4. Dan Welsby et al., "Unextractable Fossil Fuels in a 1.5°C World," *Nature*, September 2021.

5. International Energy Agency, "Record Clean Energy Spending Is Set to Help Global Energy Investment Grow by 8% in 2022," June 2022, https//www.iea.org/new/record-clean-energy-spending-is-set-to-help-global-investment-grow-by-8-in-2022,

6. Marc Davis, "U.S. Government Financial Bailouts," *Investopedia*, October 2021.

7. https://www.bostonreview.net/articles/neoliberalisms-bailout-problem. The $4 trillion can be confusing. Congress approved $5 trillion in pandemic stimulus at the beginning of 2020, with about $1.7 billion going to businesses. Pollin's $4 trillion refers to additional funds authorized by the Federal Reserve to "bail out the corporate sector in 2020" during the COVID lockdown and recession.

8. Isabella Kaminski, "Fossil Fuel Industry Faces Surge in Climate Lawsuits," *The Guardian*, June 2022.

9. "Climate Change Litigation Data Bases," Columbia University, 2022. https://climatecasechart .com/

10. https://www.dh.com/en/frances-citizen-climate-assembly-a-failed-experiment/a-56528234.

11. https://www.brennancenter.org/our-work/research-reports/citizens-united-explained.

12. Martin Gilens and Benjamin Page, "Study: Congress Literally Doesn't Care What You Think," US Represent.us, nd, https://act.represent.us/sign/problempoll-fba/.

13. Kate Kelly, Adam Playford, and Alicia Parlapiano, "Stock Trades Reported by Nearly a Fifth of Congress Show Possible Conflicts," *New York Times*, September 2022.

14. https://hbr.org/2017/02/the-growing-conflict-of-interest-problem-in-the-u-s-congress.

15. https://www.sinema.senate.gov/sinema-shaped-inflation-reduction-act-passes-senate.

16. Chad Raphael, "The FCC's Broadcast News Distortion Rules: Regulation by Drooping Eyelid," Santa Clara University Scholar Commons, Summer 2001, https://scholarcommons .scu.edu/comm/16/.

17. National Environmental Policy Act of 1969, as amended. https://www.govinfo.gov/content /pkg/STATUTE-83/pdf/STATUTE-83-pg852.pdf.

18. R. Costanza et al., "Changes in the Global Value of Ecosystem Services," *Global Environmental Change* 26 (May 2014): 152–158.

19. Todd BenDor et al., "Estimating the Size and Impact of the Ecological Restoration Economy," *PLOS One* (June17, 2015).

20. National Environmental Policy Act sec. 102.

21. https://bailoutwatch.org/overview.

22. https://www.wilsoncenter.org/solar-radiation-management.

23. https://www.ohchr.org/en/statements-amd-speeches/2022/08/crisis-and-fragility -democracy-world.

24. https://foreignpolicy.com/2022/08/06review-new-authoritarianism-spin-dictators-age -strongman-great-experiment/.

25. https://www.washingtonpost.com/opinions/2022/08/29/bucerius-global-leadership -america-democratic-decline/.

26. Dominic Packer and Jay Van Bavel, "The Myth of Tribalism," *The Atlantic*, January 2022.
27. John F. Kennedy, Speech at Rice University, September 12, 1962.

CHAPTER 9

1. Fred Krupp, "George H.W. Bush, environmental hero: He exemplified the real art of the deal," New York Daily News, December 3, 2018, https://www.nydailynews.com/opinion/ny-oped-george-h-w-bush-environmental-hero-20181203-story.html.
2. Alec Tyson and Bryan Kennedy, "Two-Thirds of Americans Think Government Should Do More on Climate," Pew Research Center, June 23, 2020, https://www.pewresearch.org/science/2020/06/23/two-thirds-of-americans-think-government-should-do-more-on-climate/.
3. "Governance" is a well-established concept in global politics that first emerged in the 1990s. In the years since, a vast literature has refined and applied the concept. See, for example, Jan Kooiman, ed., *Modern Governance: New Government-Society Interactions* (London: Sage, 1993); Lawrence S. Finkelstein, "What Is Global Governance?" *Global Governance* 1 (1995), 367–372; Oliver E. Williamson, *The Mechanisms of Governance* (New York: Oxford University Press, 1996); Jan Kooiman, *Governing as Governance* (London: Sage, 2003); Maria Carmen Lemos and Arun Agrawal, "Environmental Governance," *Annual Review of Environmental Resources* 31 (2006): 297–325.
4. James C. Scott, *Seeing Like a State: How Certain Schemes to Improve the Human Condition Have Failed* (New Haven, CT: Yale University Press, 1998).
5. Amarta Sen noted that in contrast, democracies, because they "have to win elections and face public criticism . . . have strong incentive to undertake measures to avert famines and other catastrophes." Amarta Sen, *Democracy as Freedom* (Oxford: Oxford University Press, 1999).
6. See, for example, Ann Florini et al., *Governing Carbon Markets*, report prepared for the Finance for Biodiversity Initiative, April 2022.
7. See, for example, Anne-Marie Slaughter and Gordon LaForge, "Opening Up the Order: A More Inclusive International System," *Foreign Affairs* (March/April 2021); Ann Florini, "Collaboration across Business, Government, and Civil Society as a Key Social Innovation" in *Handbook of Inclusive Innovation: The Role of Organizations, Markets and Communities in Social Innovation*, ed. Gerard George et al. (Cheltenham, UK: Edward Elgar, 2019).
8. See https://foodsystems.community/information-note-on-follow-up-to-the-un-food-systems-summit/.
9. A handful of jurisdictions are experimenting with so-called citizens assemblies, democratic bodies of citizens selected randomly, though based on demographic criteria reflective of a nation, city, or other political unit. France's Citizen's Convention on Climate (Convention Citoyenne pour le Climat) is a panel of 150 French citizens chosen by lot and empowered to propose climate-related legislation and regulation (see https://www.conventioncitoyennepourleclimat.fr/en/).
10. See reporting from, among others, Reuters and Bloomberg at https://www.reuters.com/world/china/china-slams-firms-falsifying-carbon-data-2022-03-14/ and https://www.bloomberg

.com/news/articles/2022-01-28/china-jails-almost-50-steel-executives-for-faking-emissions-data#xj4y7vzkg, as well as academic research, such as Liqun Peng et al., "Underreported Coal in Statistics: A Survey-Based Solid Fuel Consumption and Emission Inventory for the Rural Residential Sector in China," *Applied Energy* 235 (2019): 1169–1182.

11. https://www.nytimes.com/2022/04/13/business/china-covid-zero-shanghai.html.

12. See https://www.law.cornell.edu/wex/public_benefit_corporation.

13. Megan Brenan, "Americans' Trust in Media Dips to Second Lowest on Record," Gallup, October 7, 2021, https://news.gallup.com/poll/355526/americans-trust-media-dips-second-lowest-record.aspx, and "Public Trust in Government: 1958–2022," Pew Research Center, June 6, 2022, https://www.pewresearch.org/politics/2022/06/06/public-trust-in-government-1958-2022/.

14. Yuval Noah Harari, *Sapiens: A Brief History of Humankind*, New York: HarperCollins, 2015.

15. Sophie Yeo and Simon Evans, "The 35 Countries Cutting the Link between Economic Growth and Emissions," *Carbon Brief*, April 5, 2016, https://www.carbonbrief.org/the-35-countries-cutting-the-link-between-economic-growth-and-emissions/.

16. Deloitte, "The Turning Point: A New Economic Climate in the United States," January 2022, https://www2.deloitte.com/content/dam/Deloitte/us/Documents/about-deloitte/us-the-turning-point-a-new-economic-climate-in-the-united-states-january-2022.pdf.

17. Swiss Re Group, "World Economy Set to Lose Up to 18% GDP from Climate Change if No Action Taken, Reveals Swiss Re Institute's Stress-Test Analysis," April 22, 2021, https://www.swissre.com/media/press-release/nr-20210422-economics-of-climate-change-risks.html.

18. See, for example, Jacob Hay, "New Poll Shows Surge in Concern about Nature and Continued Bipartisan Support for Conservation among Western Voters," Colorado College State of the Rockies Project, February 4, 2021, https://www.coloradocollege.edu/other/stateoftherockies/conservationinthewest/2021/2021-RevisedConservation-in-the-West-Poll-National-Release.pdf.

19. See, for example, Robert D. Putnam, *Bowling Alone: The Collapse and Revival of American Community* (New York: Simon and Schuster, 2000); Theda Skocpol, Rachael V. Cobb, and Casey Andrew Klofstad, "Disconnection and Reorganization: The Transformation of Civic Life in Late-Twentieth-Century America," *Studies in American Political Development* 19 (Fall 2005): 137–156.

20. See, for example, John T. Cacioppo and Stephanie Cacioppo, "The Growing Problem of Loneliness," *The Lancet* 391, no. 10119 (February 2018): 426; Sian Leah Beilock, "Why Young Americans Are Lonely," *Scientific American*, July 27, 2020, https://www.scientificamerican.com/article/why-young-americans-are-lonely/.

CHAPTER 10

1. Arwa madawi, "On its 100th birthday in 1959, Edward Teller warned the oil industry about global warming," https://www.theguardian.com/enviornment/climate-consensus-97per-cent/2018/jan/01/on-its-hundredth-birthday-in-1959-edward-teller-warned-the-oil-industry-about-global-global-warming.

2. “Frank Capra’s Science Film *The Unchained Goddess* Warns of Climate Change in 1958,” *Open Culture*, April 8, 2015, https://www.openculture.com/2015/04/frank-capras-science-film-the-unchained-goddess-warns-of-climate-change-in-1958.html.
3. James Gustave Speth, *They Knew: The US Federal Government’s Fifty-Year Role in Causing the Climate Crisis* (Cambridge, MA: MIT Press, 2022).
4. National Inquiry on Climate Change, https://chr.gov.ph/nicc-2/.
5. Juliana v. U.S., 947 F.3d 1159, 1164–1166 (9th Cir.2020) (acknowledging that climate change is real “and occurring at an increasingly rapid pace” and auguring an “apocalypse”).
6. Jedediah Purdy, “Climate Change and the Limits of the Possible,” *Duke Environmental Law and Policy Forum* 18, no. 2 (2008): 289–306; Richard J. Lazarus, “Super Wicked Problems and Climate Change: Restraining the Present to Liberate the Future,” *Cornell Law Review* 94 (2009): 1159–1187.
7. J. B. Ruhl and Robin Kundis Craig, “4°C,” *Minnesota Law Review* 106 (2021): 192–277.
8. Marbury v. Madison, 5 U.S.137 1803.
9. M’Culloch v. State of Maryland, 17 U.S. 316 (1819).
10. United States v. Lopez, 514 U.S. 549 (1995).
11. Morrison v. Olson, 529 U.S. 598 (2000).
12. Gonzales v. Raich, 545 U.S. 1 (2005).
13. Missouri v. Holland, 252 U.S. 416 (1920).
14. South Dakota v. Dole, 483 U.S. 203 (1987).
15. Kleppe v. New Mexico, 426 U.S. 529 (1976), 539 (quoting United States v. San Francisco, 310 U.S. 16 (1940), 29).
16. Fort Leavenworth R. Company v. Lowe, 114 U.S. 525 (1885).
17. Midwest Oil Company v. United States, 236 U.S. 459 (1915).
18. New York v. United States, 505 U.S. 144 (1992).
19. Seminole Tribe of Florida v. Florida, 517 U.S. 44 (1995).
20. Alden v. Maine, 527 U.S. 706 (1999).
21. Federal Maritime Commission v. S.C. State Ports Authority, 535 U.S. 743 (2002).
22. Ex parte Young, 209 U.S. 123 (1908) (ordering a state official to comply with federal law on ground that he lacked constitutional authority to commit unconstitutional act).
23. Keystone Bituminous Coal Association v. DeBenedictis, 480 U.S. 470 (1987) (state statute restricting mining that resulted in surface subsidence was not a taking).
24. Penn Central Transport Company v. New York City, 438 U.S. 104 (1978), 124.
25. Nollan v. California Coastal Commission, 483 U.S. 825 (1987).

26. Dolan v. City of Tigard, 512 U.S. 374 (1994).
27. Lucas v. South Carolina Coastal Council, 505 U.S. 1003 (1992).
28. Palazzolo v. Rhode Island, 533 U.S. 606 (2001).
29. See, for example, Loving v. Virginia, 388 U.S. 1 (1967) (fundamental right to marriage); Meyer v. Nebraska, 262 U.S. 390 (1923) (fundamental right to childrearing).
30. See Juliana v. United States, 217 F.Supp.3d 1224, 1250 (D. Or. 2016) ("Just as marriage is the 'foundation of the family,' a stable climate system is quite literally the foundation" of society and civilization), rev'd on other grounds, 947 F.3d 1159 (9th Cir. 2020).
31. Washington v. Davis, 426 U.S. 229 (1976).
32. See, for example, McDonald v. City of Chicago, 561 U.S. 742 (2020) (Thomas, J. concurring); Timbs v. Indiana, 139 S.Ct. 682 (2019) (Gorsuch, J. concurring).
33. Slaughter-House Cases, 83 U.S. 36 (1873).
34. Whitman v. American Trucking Association, 531 U.S. 457 (2001), 473–474.
35. Baker v. Carr, 369 U.S. 186 (1962).
36. Colegrove v. Green, 328 U.S. 549 (1946), 556.
37. Massachusetts v. EPA, 549 U.S. 497 (2007).
38. American Electric Power Company v. Connecticut, 564 U.S. 410 (2011), 427–428.
39. Lujan v. National Wildlife Federation, 497 U.S. 871 (1990).
40. Sierra Club v. Morton, 405 U.S. 727 (1972).
41. Friends of Earth, Inc. v. Laidlaw Environmental Services., 528 U.S. 167 (2000).
42. Massachusetts v. EPA, 549 U.S. 497 (2007).
43. Nuclear Information & Resource Service v. Nuclear Regulatory Commission, 457 F.3d 941 (9th Cir. 2006).
44. Philadelphia v. New Jersey, 347 U.S. 617 (1978).
45. Oregon Waste Sys. v. Department of Environmental Quality, 511 U.S. 93 (1994); Fort Gratiot Sanitary Landfill v. Michigan Department of Natural Resources, 504 U.S. 353 (1992).
46. C & A Carbone, Inc. v. Clarkstown, 511 U.S. 383 (1994).
47. Sporhase v. Nebraska, 458 U.S. 941 (1982).
48. New England Power Company v. New Hampshire, 455 U.S. 331 (1982).
49. New Energy Company of Indiana v. Limbach, 486 U.S. 269 (1988).
50. Wyoming v. Oklahoma, 502 U.S. 437 (1992).
51. United Haulers Association v. Oneida-Herkimer Solid Waste Management Authority, 550 U.S. 330 (2007), 342–343.

52. Milwaukee v. Illinois, 451 U.S. 304 (1981).

53. Kimberly K. Smith, *The Conservation Constitution* (Lawrence: University Press of Kansas, 2019), 14 ("If the Constitution was meant to facilitate a national, coordinated policy on natural resource use and development, it failed. . . . The federal government . . . had virtually no explicit authority over natural resources.")

54. Lynda L. Butler, "Property's Problem with Extremes," *Wake Forest Law Review* 55, no. 1 (2020): 46–47.

55. Juliana v. U.S., 947 F.3d 1159 (9th Cir. 2020), 1171–1172 (citations omitted).

56. American Electric Power Company v. Connecticut, 564 U.S. 410, 427–428 (2011).

57. West Virginia v. EPA, No. 20–1530, 2022 WL 2347278, 18 (U.S. February 28, 2022) (citation omitted).

58. Dobbs v. Jackson Women's Health Organization, 142 S. Ct. 2228, 2242 (2022) (internal citations and quotation marks omitted).

59. James R. May and Erin Daly, "Can the U.S. Constitution Encompass a Right to a Stable Climate (Yes, It Can)," *UCLA Environmental Law Review* 39 (2021).

60. "Yale Climate Opinion Maps 2021," https://climatecommunication.yale.edu/visualizations-data/ycom-us/ (last visited June 8, 2022).

61. Pamela S. Karlan, "The New Countermajoritarian Difficulty," *California Law Review* 109, no. 6 (December 2021): 2323–2334 ("The worsening disjuncture between where Americans live and how the Constitution allocates political power is a major source of a new countermajoritarian difficulty, which lies not only in the courts, but in the Senate and the Electoral College as well. Put squarely, our political system may be incapable of reflecting the new majority").

62. Citizens United v. Federal Election Commission, 558 US 310 (2010).

63. Stephan Lewandowsky, "Climate Change Disinformation and How to Combat It," *Annual Review of Public Health* 42 (2021): 1–21; Riley E. Dunlap and Aaron M. McCright, "Challenging Climate Change: The Denial Countermovement," in *Climate Change and Society: Sociological Perspectives*, ed. Riley E. Dunlap and Robert J. Brulle (Oxford: Oxford University Press, 2015), 305–332; Geoffrey Supran and Naomi Oreskes, "Assessing ExxonMobil's Climate Change Communications (1977–2014)," *Environmental Research Letters* 10 (2017), https://iopscience.iop.org/article/10.1088/1748-9326/aa815f/pdf.

64. Ryan D. Doerfler and Samuel Moyn, "The Constitution Is Broken and Should Not Be Reclaimed," *New York Times*, August 22, 2022, https://www.nytimes.com/2022/08/19/opinion/liberals-constitution.html.

CHAPTER 12

1. Steven Pinker, *Enlightenment Now: The Case for Reason, Science, Humanism, and Progress* (New York: Viking, 2018), 200.

2. Daniel J. Fiorino, *Can Democracy Handle Climate Change?* (Cambridge, UK: Polity Press, 2018).

3. Daniel Sarewitz, *Frontiers of Illusion: Science, Technology, and the Politics of Progress* (Philadelphia: Temple University Press, 1996), 10–11.

4. David H. Guston et al., "Responsible Innovation: Motivations for a New Journal," *Journal of Responsible Innovation* 1, no. 1 (2014): 1–8.

5. E. O. Wilson, "An Intellectual Entente," *Harvard Magazine*, September 10, 2009.

6. Michael M. Crow, "None Dare Call It Hubris: The Limits of Knowledge," *Issues in Science and Technology* 23, no. 2 (Winter 2007): 29–32.

7. Braden R. Allenby and Daniel Sarewitz, *The Techno-Human Condition* (Cambridge, MA: MIT Press, 2011); David W. Orr, *Earth in Mind: On Education, Environment, and the Human Prospect* (Washington, DC: Island Press, 2004), 1–2.

8. Ronald J. Daniels, "Universities Are Shunning Their Responsibility to Democracy," *The Atlantic*, October 3, 2021.

9. Ronald J. Daniels, Grant Shreve and Phillip Spector, *What Universities Owe Democracy* (Baltimore, MD: Johns Hopkins University Press, 2021), 20, 116.

10. Suzanne Mettler, *Degrees of Inequality: How the Politics of Higher Education Sabotaged the American Dream* (New York: Basic Books, 2014), 191.

11. See http://www.malegislature.gov/laws/constitution; http://www.harvard.edu/history.

12. George Thomas, *The Founders and the Idea of a National University: Constituting the American Mind* (Cambridge, UK: Cambridge University Press, 2015); Albert Castel, "The Founding Fathers and the Vision of a National University," *History of Education Quarterly* 4, no. 4 (December 1964): 280–302.

13. John R. Thelin, *A History of American Higher Education*, 3rd ed. (Baltimore, MD: Johns Hopkins University Press, 2019), 42.

14. Thomas Jefferson to William Charles Jarvis, September 28, 1820, in *Thomas Jefferson, The Writings of Thomas Jefferson*, vol. 10, ed. Paul L. Ford (New York: G. P. Putnam's Sons, 1892–1899), 161.

15. The Morrill Act is officially designated "An Act Donating Public Lands to the Several States and Territories which may Provide Colleges for the Benefit of Agriculture and Mechanic Arts": Act of July 2, 1862, chap. 130, 12 Stat. 503, 7 U.S.C., quoted in John R. Thelin, *Essential Documents in the History of American Higher Education* (Baltimore, MD: Johns Hopkins University Press, 2014), 76–79.

16. Franklin D. Roosevelt, "State of the Union Address," January 11, 1944.

17. President's Commission on Higher Education, *Higher Education for American Democracy*, vol. 1 (New York: Harper & Brothers, 1947), 36.

18. Louis Menand, *The Marketplace of Ideas: Reform and Resistance in the American University* (New York: W. W. Norton & Company, 2010), 23.

19. Daniels, Shreve, and Spector, *What Universities Owe Democracy*, 108.

20. Harvard Committee/James Conant, *General Education in a Free Society: Report of the Harvard Committee* (Cambridge, MA: Harvard University Press, 1945), x.

21. Daniels, Shreve, and Spector, *What Universities Owe Democracy*, 110.

22. Danielle Allen et al., "Educating for American Democracy: Excellence in History and Civics for All Learners," *iCivics* (March 2021), www.educatingforamericandemocracy.org.

23. "Reimagining the AGEC for the 21st Century," *AZTransfer,* https://aztransfer.com/about/reimagining_the_agec.html.

24. Arizona Board of Regents, "Policy 2–210: General Education," https://public.azregents.edu/Policy%20Manual/2-210%20General%20Education.pdf, 1.

25. https://www.asu.edu/about/charter-mission.

26. https://thecollege.asu.edu/climate-conferencehttps://aztransfer.com/about/reimagining_the_agec.html.

27. Daniels, Shreve, and Spector, *What Universities Owe Democracy*, 242.

28. Orr, *Earth in Mind*, 114.

29. Michael M. Crow and William B. Dabars, *The Fifth Wave: The Evolution of American Higher Education* (Baltimore: Johns Hopkins University Press, 2020), 67, 214.

30. See especially Walter W. McMahon, *Higher Learning, Greater Good: The Private and Social Benefits of Higher Education* (Baltimore, MD: Johns Hopkins University Press, 2009).

31. William G. Bowen, Martin A. Kurzweil, and Eugene M. Tobin, *Equity and Excellence in American Higher Education* (Charlottesville: University of Virginia Press, 2005), 1–4.

32. Michael M. Crow and William B. Dabars, *Designing the New American University* (Baltimore, MD: Johns Hopkins University Press, 2015).

33. Crow and Dabars, *The Fifth Wave.*

34. Michael M. Crow, William B. Dabars, and Derrick M. Anderson, *Universal Learning: Democratizing American Higher Education* (Baltimore, MD: Johns Hopkins University Press, forthcoming 2023).

35. Michael M. Crow and William B. Dabars, "The Emergence of the Fifth Wave in American Higher Education," *Issues in Science and Technology* 36, no. 3 (Spring 2020): 71–74.

36. Jane Lubchenco, "Entering the Century of the Environment: A New Social Contract for Science," *Science* 279 (1998): 491–497.

37. Peter M. Vitousek et al., "Human Domination of Earth's Ecosystems," *Science* 277, no. 5325 (1997): 494.

38. Intergovernmental Panel on Climate Change (IPCC), *Global Warming of 1.5° C: An IPCC Special Report on the Impacts of Global Warming of 1.5° C above Pre-industrial Levels and Related Global Greenhouse Gas Emission Pathways*, ed. V. Masson-Delmotte et al. (Geneva: World Meteorological Organization, 2018).

39. Braden R. Allenby and Daniel Sarewitz, "We've Made a World We Cannot Control," *New Scientist* 210: 28–29.

40. Braden R. Allenby and Daniel Sarewitz, *The Techno-Human Condition* (Cambridge, MA: MIT Press, 2011).

41. Allenby and Sarewitz, *Techno-Human Condition.*

42. Jack Stilgoe, Richard Owen, and Phil Macnaghten, "Developing a Framework for Responsible Innovation," *Research Policy* 42, no. 9 (2013): 1568–1580.

43. John Parker and Beatrice Crona, "On Being All Things to All People: Boundary Organizations and the Contemporary Research University," *Social Studies of Science* 42, no. 2 (2012): 262–289.

44. William B. Dabars and Kevin T. Dwyer, "Toward Institutionalization of Responsible Innovation in the Contemporary Research University: Insights from Case Studies of Arizona State University," *Journal of Responsible Innovation* 9, no. 1 (2022): 114–123, https://doi.org/10.1080/23299460.2022.2042983.

45. John Dewey, "The Need of an Industrial Education in an Industrial Democracy," *John Dewey: The Middle Works, 1899–1924*, vol. 10, ed. Jo Ann Boydston (Carbondale: Southern Illinois University Press, 1976–1983), 139. See also Robert B. Westbrook, *John Dewey and American Democracy* (Ithaca, NY: Cornell University Press, 1991), especially chapter 6.

46. John Dewey, "Progress," in *John Dewey: The Middle Works*, vol. 9, 234–243, quoted in Philip Kitcher, "Social Progress," *Social Philosophy and Policy* 34, no. 2 (Winter 2017): 46–65.

47. Ann M. Pendleton-Jullian and John Seely Brown, *Design Unbound: Designing for Emergence in a White-Water World* (Cambridge, MA: MIT Press, 2018), vol. 1, 43.

48. Hans Jonas, *The Imperative of Responsibility: In Search of an Ethics for the Technological Age* (Chicago: University of Chicago Press, 1984).

49. Yaron Ezrahi, *Imagined Democracies: Necessary Political Fictions* (Cambridge, MA: Cambridge University Press, 2012).

50. Herbert A. Simon, *The Sciences of the Artificial*, 3rd ed. (Cambridge, MA: MIT Press, [1966] 1996).

51. W. Daniel Hillis, "The Enlightenment Is Dead, Long Live the Entanglement," *Journal of Design and Science,* February 22, 2016, https://doi.org/10.21428/1a042043.

52. Allenby and Sarewitz, *Techno-Human Condition*, 44.

53. Crow, "None Dare Call It Hubris," 29–32.

CHAPTER 13

1. https://transcripts.cnn.com/show/cnr/date/2022-08-17/segment/04.

2. John W. Powell, *Report on the Lands of the Arid Region of the United States* (Washington, DC: Government Printing Office, 1878).

3. John F. Ross, *The Promise of the Grand Canyon* (New York: Viking, 2018).

4. Eric Kuhn and John Fleck, *Science Be Dammed* (Tucson: University of Arizona Press, 2019).

5. Daniel Rothberg, "The Coming Crisis along the Colorado River," *New York Times*, August 4, 2022, sec. SR, 8, https://www.nytimes.com/2022/08/04/opinion/drought-climate-colorado-river.html.

6. https://www.army.mil/article/198095/dwight_d_eisenhower_and_the_birth_of_the_interstate_highway_system.

7. Lewis Mumford, *The Highway and the City* (Westport, CT: Praeger, 1963).

8. Robert D. Yaro, Ming Zhang, and Frederick R. Steiner, *Megaregions and America's Future* (Cambridge, MA: Lincoln Institute of Land Policy, 2022).

9. https://10across.com/.

10. Edward Luce, *Time to Start Thinking* (New York: Atlantic Monthly Press, 2012).

11. Oskar J. W. Hansen, "With the Look of Eagles. The Sculptures at Boulder Dam—Part I," The Reclamation Era, vol. 32 (Water and Power Resources Service, U.S. Department of the Interior, February 1942), 30. https://books.google.com/books?id=2toUAKvxoVoC&pg=PA30&lpg=PA30&dq=%22a+monument+to+collective+genius+exerting+itself+in+community+efforts+around+a+common+need+or+ideal.%22&source=bl&ots=1uvPcH0rgm&sig=ACfU3U0PP4yVMg1fQMEZ7wy6jpRaGF94pQ&hl=en&sa=X&ved=2ahUKEwiz747HteP8AhXEL0QIHXWoDMQQ6AF6BAgwEAM#v=onepage&q=%22a%20monument%20to%20collective%20genius%20exerting%20itself%20in%20community%20efforts%20around%20a%20common%20need%20or%20ideal.%22&f=false

CHAPTER 14

1. For summaries of more than a thousand studies, see the Children & Nature Network Research Library.

2. https://pubmed.ncbi.nim.nih.gov/30873068/.

3. http://colorado.construction.com/ddj/archive/2009/090518_ddj4.asp.

Contributors

William J. Barber III grew up in eastern North Carolina, where, under the tutelage of his father, Bishop William J. Barber II (of the Poor People's Campaign), and mother, Rebecca Barber, he developed at an early age a deep commitment to social justice and environmental stewardship. William now works as an environmental and climate justice scholar and advocate, with nearly a decade of social justice organizing experience and deep academic training in both the science and the law behind environmental and climate issues. William is the founder and CEO of the Rural Beacon Initiative—a business focused on strategies to increase the participation of BIPOC leaders in the energy and food supply chains. His work has been featured in PayPal's Rising Leaders Series, as well as in NextGen America's 30 Around 30: Young Black Changemakers, working to build a more just and equitable society.

William S. Becker is the executive director of the Presidential Climate Action Project, founded in 2007 to develop policy recommendations for US presidents to address global climate change and the transition to clean energy. Prior to 2007, he was a senior official at the US Department of Energy, where he specialized in renewable energy and helping flood-prone communities adapt to the growing impacts of climate change.

Holly Jean Buck is a geographer and social scientist who researches public engagement with emerging climate technologies. She holds a PhD in development sociology from Cornell University and is an assistant professor of environment and sustainability at the University at Buffalo in Buffalo, New York. She is the author of *Ending Fossil Fuels: Why Net Zero Is Not Enough* and *After Geoengineering*.

Stan Cox is a plant geneticist for thirty-three years working with the US Department of Agriculture and the Land Institute in Salina, Kansas. He currently serves as a research fellow in ecosphere studies at the Land Institute. He is the author of six books, most recently *The Path to a Livable Future* (City Lights, 2021), *The Green New Deal and Beyond* (City Lights, 2020), and, with Paul Cox, *How the World Breaks: Life in Catastrophe's Path, from the Caribbean to Siberia* (The New Press, 2016).

Michael M. Crow is president of Arizona State University, Arizona State University Foundation Leadership Chair, and Professor of Science and Technology Policy. He was previously executive vice provost and professor of science and technology policy at Columbia University. Crow and Barry Bozeman are coauthors of *Public Values Leadership: Striving to Achieve Democratic Ideals* (Johns Hopkins University Press, 2021). Crow and William Dabars are coauthors of *Designing the New American University* (Johns Hopkins University Press, 2015) and *The Fifth Wave: The Evolution of American Higher Education* (Johns Hopkins University Press, 2020).

William B. Dabars is a research professor in the School for the Future of Innovation in Society, Senior Global Futures Scholar in the Julie Ann Wrigley Global Futures Laboratory, and senior director of research for the New American University in the Office of the President, Arizona State University. He and Michael M. Crow are coauthors of *Designing the New American University* (Johns Hopkins University Press, 2015) and *The Fifth Wave: The Evolution of American Higher Education* (Johns Hopkins University Press, 2020).

Ann Florini is a Fellow in the Political Reform Program at New America, working on how innovative governance tools can help address the intertwined challenges of climate change and democratic decay. She is also senior adviser to NatureFinance and the Task Force on Nature Markets; a Senior Global Futures Scientist at the Julie Ann Wrigley Global Futures Lab and a Professor of Practice at the Thunderbird School of Global Management, Arizona State University; a founding board member of the Economics of Mutuality Foundation; and a founding member of the Council on Economic Policies. She previously taught at the National University of Singapore and held senior appointments at the Brookings Institution and the Carnegie Endowment for International Peace.

David H. Guston is Foundation Professor in Arizona State University's School for the Future of Innovation in Society and Associate Vice Provost for Discovery, Engagement and Outcomes in ASU's Julie Ann Wrigley Global Futures Laboratory. His work defining and articulating such crucial concepts as boundary organizations, responsible innovation, and anticipatory governance has led to their incorporation into governmental and nongovernmental organizations and research programs in many parts of the world. He is the author of award-winning books: *Between Politics and Science* (2000) and coeditor of the bicentennial edition of Mary Shelley's *Frankenstein: Annotated for Scientists, Engineers, and Creators of All Kinds.* He has a bachelor's degree from Yale and a PhD in political science from MIT. He is an elected fellow of the American Association for the Advancement of Science.

Katrina Fischer Kuh is the Haub Distinguished Professor of Environmental Law at the Elisabeth Haub School of Law at Pace University and coeditor of *Climate Change Law: An Introduction* and *The Law of Adaptation to Climate Change: United States and International Aspects*. She is also a member of the Environmental Law Collaborative and serves on the board of Green Amendments for the Generations.

Gordon LaForge is a senior policy analyst at New America, working on global governance, especially in the areas of environment and technology. He is a lecturer at the Thunderbird School of Global Management, Arizona State University, where he teaches on international affairs and network theory. He was previously senior researcher at Princeton University's Innovations for Successful Societies program, a geopolitical analyst at the predictive analytics firm Predata, and a journalist in Indonesia, where he was awarded two Fulbright fellowships.

Hélène Landemore is a professor of political science at Yale University and holds a Ph.D. from Harvard and a Master's degree from the Sorbonne. She is the author, most recently, of *Open Democracy*.

Frances Moore Lappé is the author or coauthor of twenty books and cofounder of the Small Planet Institute in Cambridge, Massachusetts. Her first book, *Diet for a Small Planet* (1971), has sold more than three million copies. Many of her books focus on themes of "living democracy"—suggesting not only a government accountable to citizens but a way of living aligned with the deep human need for connection, meaning, and power. She is the recipient of twenty honorary doctorates from distinguished institutions, most recently Indiana University. In 1987 she received the Right Livelihood Award, often called the Alternative Nobel. The fiftieth-anniversary edition of *Diet for a Small Planet* was released in 2021, with features in the *New York Times*, *Boston Globe*, and other major outlets. Her 2019 *New York Times* interview began, "Frances Moore Lappé changed how we eat. She wants to do the same for our democracy."

Daniel Lindvall has a PhD in sociology and is a researcher at the Climate Change Leadership Initiative at the Department of Earth Sciences in Uppsala University. He has written several books on the issues of climate change and democracy and has for several years worked on democratic affairs for the Swedish government as well as for international organizations. He was the principle inquiry secretary for the Swedish government's Democracy Commission in 2014–2016.

Richard Louv is the author of ten books, including *Our Wild Calling*, *Vitamin N*, *The Nature Principle*, and *Last Child in the Woods*, from which parts of this piece were adapted. Louv is cofounder of the Children & Nature Network (C&NN). For information about recent research on the human connection to the natural world, see C&NN's online Research Library.

Frederick W. Mayer is dean of the Josef Korbel School of International Studies at the University of Denver. Previously, he served on the faculty at Duke University for many years, where he was the founder and first director of the Center for Political Innovation, Service, and Leadership (POLIS). He has published on a range of topics, including negotiation and collective action, symbolic and narrative politics, international trade, global value chains, private governance, and environmental communication. His most recent book,

Narrative Politics: Stories and Collective Action (Oxford University Press, 2014) explores the power of shared narrative in solving collective action problems.

James R. May, Esq. is Distinguished Professor of Law at Widener University Delaware Law School and the Haub Visiting Scholar at the Elizabeth Haub School of Law at Pace University. May is a former federal trial and appellate litigator and one of the world's leading authorities on environmental human rights.

Bill McKibben is a contributing writer to *The New Yorker*, and a founder of Third Act, which organizes people over the age of 60 to work on climate and racial justice. He founded the first global grassroots climate campaign, 350.org, and serves as the Schumann Distinguished Professor in Residence at Middlebury College in Vermont. In 2014 he was awarded the Right Livelihood Prize, sometimes called the "alternative Nobel," in the Swedish Parliament. He's also won the Gandhi Peace Award, and honorary degrees from 19 colleges and universities. He has written more than a dozen books about the environment, including his first, *The End of Nature*, published in 1989, and his latest book is *The Flag, the Cross, and the Station Wagon: A Graying American Looks Back at his Suburban Boyhood and Wonders What the Hell Happened.*

Michael Oppenheimer is the Albert G. Milbank Professor of Geosciences and International Affairs at Princeton University and Director of its Center for Policy Research on Energy and the Environment. Oppenheimer has been an author of reports of the Intergovernmental Panel on Climate Change (IPCC), which won the Nobel Peace Prize in 2007, since its First Assessment Report (1990). He is a science advisor to the Environmental Defense Fund and member of several boards of directors including the Board of the Trust for Governors Island (NYC), the future site of a major climate science research and education center focused on solutions to this problem. He is a Heinz Award winner and a Fellow of the American Association for the Advancement of Science. Much of his work has centered on defining the concept of "dangerous" climate change, a key aspect of the UN Framework Convention on Climate Change and the Paris Agreement.

David W. Orr is the Paul Sears Distinguished Professor, Emeritus, from Oberlin College and a Professor of Practice at Arizona State University. He is the author of eight books and editor or coeditor of five others. He has been awarded nine honorary degrees and lifetime achievement awards from the US Green Building Council and the North American Association for Environmental Education.

Wellington "Duke" Reiter, FAIA, is the founder and director of Ten Across (10X), a nationally recognized initiative on the frontlines of demographic, economic, political, and climate change. He also serves as the senior adviser to the president of Arizona State University and as the executive director of the University City Exchange. Reiter is a Fellow of the American Institute of Architects (AIA) and a trustee of the Urban Land Institute (ULI). He is the former dean of the College of Design at ASU, the past president of the School of the Art Institute of Chicago, and a longtime faculty member in the MIT Department of Architecture.

Kim Stanley Robinson is an American science fiction writer. He is the author of twenty books, including the internationally bestselling Mars trilogy and, more recently, *Red Moon, New York 2140, and The Ministry for the Future.* He was part of the US National Science Foundation's Antarctic Artists and Writers' Program in 1995 and 2016 and a featured speaker at COP26 in Glasgow, as a guest of the UK government and the UN. His work has been translated into twenty-six languages and won awards including the Hugo, Nebula, and World Fantasy awards. In 2016, asteroid 72432 was named Kimrobinson.

Anne-Marie Slaughter is the CEO of New America and the Bert G. Kerstetter '66 University Professor Emerita of Politics and International Affairs at Princeton University. From 2009 to 2011 she served as the director of policy planning for the United States Department of State. Prior to her government service, Dr. Slaughter was the dean of Princeton's School of Public and International Affairs. She has written or edited eight books, including *The Chessboard and the Web: Strategies of Connection in a Networked World.*

Index